D1046658

TRANSISTOR–TRANSISTOR LOGIC AND ITS INTERCONNECTIONS

The Marconi Series
Covering Advances in Radio and Radar

Series Editor:
P. S. Brandon
Manager, Research Division
The Marconi Company Limited

G. E. Beck—*Navigation Systems: A Survey of Modern Electronic Aids*
M. A. R. Gunston—*Microwave Transmission-line Impedance Data*
J. A. Scarlett—*Printed Circuit Boards for Microelectronics*
J. A. Scarlett—*Transistor–Transistor Logic and its Interconnections*
J. K. Skwirzynski—*Design Theory and Data for Electronic Filters*
V. O. Stokes—*Radio Transmitters: R. F. Power Amplification*
R. A. Waldron—*Ferrites: An Introduction for Microwave Engineers*
R. A. Waldron—*Theory of Guided Electromagnetic Waves*
D. W. Watson and H. E. Wright—*Radio Direction-Finding*

Transistor–Transistor Logic and its Interconnections

A Practical Guide to Microelectronic Circuits

J. A. SCARLETT
Marconi Radar Systems Limited
Chelmsford, Essex

VAN NOSTRAND REINHOLD COMPANY
LONDON
NEW YORK CINCINNATI TORONTO MELBOURNE

VAN NOSTRAND REINHOLD COMPANY LTD
Windsor House, 46 Victoria Street, London SW1

INTERNATIONAL OFFICES
NEW YORK CINCINNATI TORONTO MELBOURNE

Library of Congress Catalog Card No. 74–171376
ISBN 0 442 07370 4

First Published 1972

Printed in Great Britain by
BUTLER AND TANNER LTD, FROME AND LONDON

Preface

This book is about T.T.L.—Transistor–Transistor Logic—and its interconnections. It is written as a practical guide to all who use, work with, or intend to use T.T.L.: systems designers, circuit designers, design engineers, commissioning and test engineers, field service engineers, and particularly students and others entering the electronics industry and those who are contemplating using integrated circuits for the first time. The aim throughout is to offer to all readers a better understanding of what T.T.L. is, how it works and how it can best be interconnected, and to offer to practical engineers some solutions to some problems and guidance in the solving of other problems.

The book is not intended to replace manufacturers' data sheets and catalogues, but to enable a reader to understand and compare the information offered by the many manufacturers of T.T.L.

Knowledge of basic transistor theory, a sound basis in the fundamentals of electronic engineering, and some knowledge of basic logic design are assumed. As far as possible a mathematical approach has been avoided, as has the use of terminology which is not in general use throughout the electronics industry. I have tried to write in plain language which can be understood with the minimum of effort by all readers, whatever their background. Generally the style adopted and the presentation of information is that which has been found to be the best way of explaining the intricacies of T.T.L. to many of the students and junior engineers who joined the Marconi Company, and in particular Computer Division, in the past four years.

Some relevant topics, such as the manufacture of integrated circuits and packaging of chips, are described very briefly indeed. I have tried to include only those facts which are essential to a full understanding of how T.T.L. works, and I have recommended other books to those who wish to know more about these particular topics.

Other topics which might be considered relevant, such as cooling, the mechanical design of boards to withstand shock and vibration, highway drivers and receivers for balanced highways, and discrete component circuits, have not been described at all, because they are topics which can not be condensed without loss of essential information. I have therefore left it to readers to use their own preferred sources of information on these topics. Also I have not attempted to include any applications of T.T.L. in this book. The manufacturers of T.T.L. issue application reports which suggest uses for the devices, and logic design is a topic which needs a full book to itself.

One other topic which must be regarded as an essential part of good utilization of T.T.L. has had to be excluded so as to keep this book down to a reasonable length. That topic is the design and manufacture of printed circuit boards, which is a topic which requires a complete book to itself—such as *Printed Circuit Boards for Microelectronics*, J. A. Scarlett, also published by Van Nostrand Reinhold (1970).

Yet another omission is that I do not recommend a 'best buy' in T.T.L. There can be a 'best buy' only when all the circumstances of an individual design are known and test facilities, etc. are considered. I hope I have written enough for readers to decide which family of T.T.L. will best suit their particular requirements. Then they must choose a supplier or suppliers for themselves, remembering that the firm which quotes the lowest initial price may not prove the cheapest in the long run—especially if high reliability, short delivery time, or a high A.Q.L. are required—and also remembering that the job of a sales representative is to sell his employer's product.

Prices are not quoted in this book. At the time of writing the semiconductor industry is fighting a keen price war, and by the time this book is printed today's prices could be very misleading. The price war and a general recession in the electronics industry can also affect the lists of firms who manufacture T.T.L. Since the bulk of this book was written, one firm, Sylvania, has ceased to manufacture T.T.L., and Motorola have taken over their product line. By the time this book is published other firms might have decided to withdraw, so the lists of manufacturers quoted here should be regarded as being of historical significance only. The logic families themselves are unlikely to be affected—most of them are second and third sourced. I have tried to make the lists of manufacturers as complete as possible, but there may be some of whom I have not heard, but who make T.T.L. devices which are compatible with the families I have included.

This book is written in sections which are, as far as possible, complete in themselves, so that an engineer with a specific query can find his answer without having to follow up an excessive number of cross-references. This has led to some necessary duplication of fundamental points, but I hope that readers who go through the book will excuse this and trust that they will not find it excessive.

For the benefit of novices with no previous knowledge of T.T.L. I have endeavoured to order the book such that in no case does the understanding of any chapter depend on knowledge which appears in a later chapter. This has led to the discussion on the current spike and its effects being split between Chapters 6 and 16.

Chapter 13 may be considered out of place. If it is regarded solely as an example of a flip-flop circuit, then naturally it should have preceded the chapter on M.S.I. However, it is included not as an example of a flip-flop but as a guide to the type of investigation and understanding which is necessary if the best use is to be got from complex devices and M.S.I. functions. As such, it concludes the portion of the book which actually describes T.T.L. devices. The particular circuit was chosen for Chapter 13 because it offered several unusual features, and the unique base-drive circuit in the clock pulse line.

The symbols and definitions used in this book may be questioned by theorists. As far as possible I have followed the recommendations in the latest B.S. specifications (such as BS 9400), but if these differed significantly from the conventions adopted by the manufacturers in their data sheets, then I have followed the conventions used in the data sheets, because practical engineers who use this book will also be using the data sheets. Thus throughout I have used 'Dee' symbols for AND gates, shields for ORs, and a small circle to indicate inversion; and unless otherwise stated, I have used positive logic convention throughout—i.e. a '1' is a high level and a '0' is a low level.

This book had its origins in the intensive investigation into T.T.L. and its interconnections which was carried out as part of my work in Computer Division, the Marconi Company, Chelmsford. Changes in the structure of the company resulted in some investigations, particularly those into long-term reliability and earthing, being curtailed, and this book can not, as a result, be as comprehensive as would have been desired. In many cases it has been possible to add to the original Computer Division work the results of work carried out in other divisions of the Marconi Company, and further material has been gathered from as wide a variety of sources as possible.

No full list of references is offered. Information has been gathered from far too wide a variety of sources. The start was inevitably the information published by the various manufacturers of T.T.L., then followed innumerable discussions with technical representatives, applications engineers, quality managers, and the designers of integrated circuits, all of which led to a better understanding of T.T.L. Perusal of as much other literature as could be obtained, attendance at conferences and symposia, and, most important of all, countless discussions with other engineers using T.T.L. completed the external sources of information. Not only does this wide variety of sources make it impossible to provide a useful list of references, it also makes it impossible to list the names of all those who have helped me in some way in my work on T.T.L., and I must acknowledge my indebtedness to the technical representatives and other staff of all the manufacturers mentioned in this book, and to the innumerable other engineers I have met in the course of my work. In particular I would like to express my thanks to the manufacturers and users of T.T.L. in the U.S.A. who made me so welcome on my recent tour of their country, and who provided so much useful material for this book. In expressing my thanks to all who helped in the direct investigations at Kensal House and at Witham, I hope that the many people involved will not be offended if I single out two names only for mention: Mr. A. M. King for his guidance on the investigations into the T.T.L. circuit, and Dr. O. Mazzariol for his work on cross-talk and ringing which provided a firm foundation for the subsequent work which led to the writing of Chapters 14 and 15. Last, but by no means least, I must thank my wife for typing the manuscript, and the staff of the Van Nostrand Reinhold Company for their advice and help.

Contents

PREFACE v

1 INTRODUCTION 1
- 1.1 T.T.L. What it is 1
- 1.2 Brief History 3
- 1.3 Alternatives to T.T.L. 4
- 1.4 Manufacturers' Literature 4
- 1.5 Use and misuse of T.T.L. 5

2 MANUFACTURE, TESTING AND RELIABILITY 6
- 2.1 Manufacture of T.T.L. 6
- 2.2 Packaging of T.T.L. 7
  - 2.2.1 'Face-up' packaging 7
    - 2.2.1.1 Separation of chips 7
    - 2.2.1.2 Packages 8
    - 2.2.1.3 Mounting and bonding 9
    - 2.2.1.4 Sealing and encapsulating 9
  - 2.2.2 'Face-down' packaging 9
    - 2.2.2.1 'Flip-chip' bonding 9
    - 2.2.2.2 'Beam lead' bonding 9
  - 2.2.3 Hybrid L.S.I. 10
- 2.3 Testing of T.T.L. 11
  - 2.3.1 Probe testing 11
  - 2.3.2 Characterization and Q.A. tests 11
- 2.4 Quality of T.T.L. 16
  - 2.4.1 Manufacturing defects 16
  - 2.4.2 Burn-in 17
  - 2.4.3 'Goods-inwards' testing 17
- 2.5 Reliability of T.T.L. 18
  - 2.5.1 Moisture and plastic packages 18
  - 2.5.2 'Purple plague' 18
  - 2.5.3 Metal migration 19
  - 2.5.4 Junction temperatures 21
  - 2.5.5 Thermal gradients 22

3 THE BASIC T.T.L. CIRCUIT 23

3.1 Circuit description and logical operation 23
3.1.1 The input 'AND' function 23
3.1.2 The phase-splitter or 'OR' function 24
3.1.3 The output stage 25
3.1.4 Capacity and parasitic diodes 26
3.2 Family differences in the basic gate 27
3.2.1 The S.U.H.L. 1 gate 28
3.2.2 The Series 54/74 gate 30
3.2.3 The Series 9000 gate 31
3.2.4 The S.U.H.L. 2 gate 32
3.2.5 The H.L.T.T.L. 2 gate 33
3.2.6 The 54H/74H gate 33
3.2.7 The RAY 3 gate 34
3.2.8 The M.T.T.L. 3 gate 34
3.2.9 The Series 54L/74L gate 34
3.2.10 The 54S/74S gate 35

4 DC AND LOW-FREQUENCY PARAMETERS 37

4.1 Input characteristic 37
4.2 Output characteristic 40
4.2.1 Low level characteristic 41
4.2.2 High level characteristic 42
4.3 Transfer characteristic 43

5 SWITCHING OF THE T.T.L. GATE 47

5.1 Simplified switching action—turn-on 47
5.2 Simplified switching action—turn-off 49
5.3 General effects of switching 51
5.3.1 Change of propagation delay with frequency 51
5.3.2 Change of device dissipation with frequency 51
5.3.3 Effects caused by the switching of two gates on the same chip 51

6 THE T.T.L. SWITCHING SPIKE 53

6.1 The current spike 53
6.1.1 Introduction 53
6.1.2 Generation of the current spike during turn-off 53
6.1.3 Generation of the current spike during turn-on 55

6.2 Effects caused by the current spike 55
6.2.1 Supply rail voltage spike 55
6.2.2 Interference on signal lines 55

7 VARIANTS OF THE BASIC T.T.L. CIRCUITS 56

7.1 The non-inverting gate or 'AND' gate 56
7.2 The 'NOR' gate 59
7.3 The 'OR' gate 59
7.4 Expanders 59
7.4.1 Correct use of expanders 60
7.4.2 'OR' expansion 60
7.4.3 'AND' expansion 61
7.4.4 Non-inverting 'OR' expansion 62
7.5 Line drivers or buffers 62
7.6 Lamp drivers 63

8 RANGE OF GATES WITHIN THE FAMILIES 65

8.1 Nomenclature 65
8.1.1 Introduction 65
8.1.2 Manufacturers' identification 65
8.1.2.1 Manufacturers' names 65
8.1.2.2 Type numbers 65
8.1.3 Simplified numbering system 66
8.1.3.1 Basic numbering system 66
8.1.3.2 Flip-flops and complex functions 67
8.1.3.3 Prefix and suffix 67
8.1.3.4 Applications to computer-aided design 68
8.2 Gate types 68
8.2.1 'NAND' Gates 68
8.2.2 'AND-OR-INVERT' gates 69
8.2.3 'AND' or non-inverting gates 72
8.2.4 Expanders 72
8.2.4.1 'AND' expanders 72
8.2.4.2 'OR' expanders 72
8.2.4.3 Non-inverting expanders 72
8.2.5 Buffers, Lamp drivers, and other types 72

9 MAIN PARAMETERS OF T.T.L. GATES 73

9.1.1 Introduction 73
9.1.2 Manufacturers' specifications 80

9.2 Speed 81
9.2.1 General 81
9.2.2 Specified figures 81
9.2.3 Typical figures 82
9.2.4 Ratio of edge speed to propagation delay 83
9.2.5 Unrequired inputs on 'NAND' gates 83
9.3 Fan-out 84
9.3.1 Introduction 84
9.3.2 Parallel connections to inputs of one gate 84
9.3.3 Specified figures 84
9.3.4 Fan-out table 88
9.3.5 Measured values 88
9.4 Noise immunity 89
9.4.1 Introduction 89
9.4.2 Specified voltage noise margins 90
9.4.2.1 '1' level 90
9.4.2.2 '0' level 90
9.4.3 Practical noise considerations 90
9.4.4 Measured low frequency parameters 91
9.4.5 High frequency transfer characteristics 91
9.5 Dissipation 96
9.5.1 Specified and typical values 96
9.5.2 Board or unit dissipation 97
9.5.3 Current surges 97
9.5.4 Unused gates 99

10 ENVIRONMENTAL VARIATIONS IN PARAMETERS 100

10.1 Temperature 100
10.1.1 Introduction 100
10.1.2 Effect of temperature on speed 100
10.1.3 Effect of temperature on fan-out 102
10.1.4 Effect of temperature on noise immunity 102
10.1.5 Effect of temperature on dissipation 103
10.1.6 Effect of temperature on the switching spike 103
10.2 Capacitance 104
10.2.1 Introduction 104
10.2.2 Effect of capacitance on speed 104
10.2.3 Effect of capacitance on fan-out 104
10.2.4 Effect of capacitance on noise immunity 104
10.2.5 Effect of capacitance on dissipation 104
10.2.6 Effect of capacitance on the switching spike 106

10.3 Supply rail variation 107
10.3.1 Introduction 107
10.3.2 Effect of rail variation on speed 107
10.3.3 Effect of rail variation on fan-out 107
10.3.3.1 Output of the driving gate 107
10.3.3.2 Input of the driving gate 107
10.3.4 Effect of rail variation on noise immunity 109
10.3.5 Effect of rail variation on dissipation 109
10.3.6 Effect of rail variation on the switching spike 110
10.4 Fan-out 110
10.4.1 Introduction 110
10.4.2 Effect of fan-out on speed 110
10.4.3 Effect of fan-out on fan-out 110
10.4.4 Effect of fan-out on noise immunity 110
10.4.5 Effect of fan-out on dissipation 110
10.4.6 Effect of fan-out on the switching spike 111
10.5 Expanders 111
10.5.1 Introduction 111
10.5.2 Effects of expanders on speed 111
10.5.2.1 'OR' expanders on inverting gates 111
10.5.2.2 'OR' expanders on non-inverting gates 112
10.5.2.3 'AND' expanders on eight-input gate 112
10.5.3 Effects of expanders on fan-out 114
10.5.4 Effects of expanders on noise immunity 114
10.5.4.1 'OR' expanders on inverting gates 114
10.5.4.2 'OR' expanders on non-inverting gates 115
10.5.4.3 'AND' expanders on eight-input gate 115
10.5.5 Effects of expanders on dissipation 115
10.5.6 Effects of expanders on the switching spike 115

11 FLIP-FLOPS 116

11.1 Types of flip-flops 116
11.1.1 R.S. flip-flops 116
11.1.2 M.S. flip-flops 116
11.1.3 J.K. flip-flops 117
11.1.4 D-Type flip-flops 117
11.1.5 types available in T.T.L. 117
11.2 Working of flip-flops 117
11.2.1 Direct coupled flip-flops 118
11.2.1.1 Direct coupled master–slave J.K. flip-flops 118
11.2.1.2 Transistor coupled master–slave J.K. flip-flops 120

11.2.1.3 Edge-triggered J.K. flip-flops 120
11.2.1.4 Edge-triggered D-type flip-flops 121
11.2.2 A.C. Coupled (charge storage) flip-flops 122
11.3 Use of flip-flops 124
11.3.1 Input loadings 124
11.3.2 Unused half packages 124
11.3.3 Clock edge speeds 125

12 MEDIUM-SCALE INTEGRATION 126

12.1 What is M.S.I.? 126
12.2 Advantages and disadvantages of M.S.I. 126
12.3 M.S.I. families 127
12.4 Range of M.S.I. devices 130
12.4.1 Registers 131
12.4.2 Latches 131
12.4.3 Encoders, decoders, and multiplexers 131
12.4.4 Counters 131
12.4.5 Arithmetic units, adders, and comparators 132
12.4.6 Memories 132

13 DETAILED CONSIDERATION OF A FLIP-FLOP CIRCUIT 133

13.1 Introduction 133
13.2 Logical operation 135
13.2.1 General 135
13.2.2 Logical state with low clock input 135
13.2.3 Logical state with high clock input 136
13.2.4 Logical operation with rising clock input 138
13.2.5 Logical operation with falling clock input 139
13.2.6 Logical operation of hold element 139
13.3 Electrical operation 139
13.3.1 General 139
13.3.2 Operation of the conventional elements 140
13.3.3 Electrical operation of elements c and e (the clock generator) 140
13.3.3.1 General 140
13.3.3.2 Input low 140
13.3.3.3 Input high 144
13.3.3.4 Input rising from low to high 144
13.3.3.5 Input falling from high to low 145
13.3.4 Electrical operation of elements f, g, and j (the locking ring) 146

13.4 Logical use of the clock or hold inputs 148
13.5 Typical electrical characteristics of the D.F.F. 148
13.5.1 General 148
13.5.2 Timings 149
13.6 Clock line driving and noise immunity 150
13.7 Chip layouts of the D-type flip-flop 150
13.7.1 Introduction 150
13.7.2 Sylvania layout 151
13.7.3 Transitron layout 153

14 THEORY OF TRANSMISSION LINE EFFECTS ON PRINTED CIRCUIT BOARDS 156

14.1 Introduction 156
14.2 Transmission line theory 157
14.2.1 Properties of a single transmission line 157
14.2.1.1 Characteristic impedance 157
14.2.1.2 Propagation delay 157
14.2.2 Reflections in a single transmission line 158
14.2.2.1 Terminations 158
14.2.2.2 Determination of reflections 160
14.2.2.3 Graphical solution of reflections 161
14.2.2.4 Critical length 162
14.2.3 Discontinuities 165
14.2.3.1 Changes in track width 165
14.2.3.2 Passive resistive discontinuities 167
14.2.3.3 Active resistive discontinuities 167
14.2.3.4 Capacitive discontinuities 167
14.2.4 The 'wired-OR' 167
14.3 The T.T.L. line 169
14.3.1 Reflections 169
14.3.1.1 Long line 169
14.3.1.2 Critical line 169
14.3.1.3 Short line 172
14.3.2 Amplitude of steps 172
14.3.2.1 Turn-off 172
14.3.2.2 Turn-on 175
14.4 Cross-talk 177
14.4.1 General case 177
14.4.1.1 Back cross-talk 177
14.4.1.2 Forward cross-talk 178
14.4.2 Effect of pick-up line terminations on back cross-talk 180

14.4.3 Cross-talk between T.T.L. lines 182
14.4.3.1 Pick-up line low conditions 183
14.4.3.2 Pick-up line high conditions 184
14.4.3.3 Pick-up lines shorter than the critical length 186
14.4.3.4 Effects of ring on the signal line 186

15 CROSS-TALK AND RING ON PRACTICAL PRINTED CIRCUIT BOARDS 187

15.1 Divergencies from microstrip lines 187
15.2 Characteristic impedance and reflections on double-sided boards 187
15.2.1 Experiments on *X–Y* co-ordinate boards 187
15.2.2 Experiments on boards with parallel tracks on both sides 189
15.3 Cross-talk in multiple lines configurations–experiments 191
15.3.1 Co-relation between experimental results and theory 191
15.3.2 Nomenclature used in multiple line configurations 191
15.4 Cross-talk in multiple lines 193
15.4.1 Effect of non-switching lines adjacent to signal and pick-up lines 193
15.4.2 Effect of multiple signal tracks switched simultaneously 193
15.4.3 Effect of signal tracks on both sides of the board 194
15.4.4 Effect of introducing earth lines either side of the pick-up line 194
15.4.5 Effect of unswitched tracks between signal and pick-up lines 195
15.5 General notes on cross-talk on double-sided boards 198
15.5.1 Logical state of back wiring 198
15.6 Cross-talk in discrete wiring 198
15.7 Critical length of pick-up lines 199

16 PRINTED CIRCUIT BOARD DESIGN FOR T.T.L. 200

16.1 Mounting of T.T.L. devices 200
16.1.1 Board layout 200
16.1.2 Track widths and characteristic impedance 201
16.1.3 Maximum track length to guarantee freedom from spurious switching caused by cross-talk 201
16.1.3.1 Noise margins 201
16.1.3.2 Voltage swing 202
16.1.3.3 Values of K_B 202

16.1.3.4 Cross-talk on double-sided boards 202
16.1.3.5 Edge speeds of T.T.L. and track length 202
16.1.3.6 Effect of reducing clearance between tracks 203
16.1.4 Other limits on track length 203
16.1.5 Application of limits to track length 204
16.2 Power distribution and voltage spike 204
16.2.1 Supply rail impedance 204
16.2.1.1 Power supply on double-sided boards 205
16.2.1.2 Calculated supply rail impedance 205
16.2.2 Voltage spikes on typical boards 207
16.2.2.1 Maximum amplitude of current spike 207
16.2.2.2 Worst-case voltage spike 208
16.2.3 Effect of the voltage spike on the output signal 209
16.2.4 Addition of voltage spikes caused by gates switching simultaneously 212
16.2.5 Effect of earth rail impedance and spikes on the board earth 212
16.3 Decoupling 213
16.3.1 Introduction 213
16.3.2 The use of internal power and earth planes 213
16.3.3 Practical decoupling on double-sided boards 215

17 PRACTICAL CONSIDERATIONS IN THE USE OF T.T.L. 217
17.1 Power supply 217
17.1.1 Normal limits 217
17.1.2 Performance while supply rail is rising or falling 217
17.2 Connections to the outputs of T.T.L. gates 218
17.2.1 Interconnection of gate outputs 218
17.2.2 Connection of gates in parallel to drive high fan-out nodes 218
17.2.3 Shorting of inputs and outputs during testing 219
17.2.4 Checking and testing T.T.L. devices on boards 220
17.2.4.1 Open circuit faults 220
17.2.4.2 Short-circuit faults 220
17.2.4.3 Parametric degradations 221
17.3 Line driving and terminations 222
17.3.1 Shunt terminating networks 222
17.3.2 Partial termination of lines driven by standard T.T.L. gates 223
17.4 Clock skew 223
17.4.1 Introduction 223
17.4.2 Avoiding clock skew 223
17.4.3 Clock distribution around a system 225

17.5 Current hogging 225
17.5.1 Strobing of long lines 225
17.5.2 Matrix driving 226
17.6 External signals to T.T.L. circuits 226
17.6.1 Slow-edged pulses 226
17.6.2 Switches 227

18 THE INFLUENCE OF T.T.L. ON SYSTEM DESIGN 228

18.1 Design of the logic system 228
18.1.1 The influence of M.S.I. on system design 229
18.2 Implementation of the system 229
18.2.1 Sub-systems 229
18.2.2 Division of the system into printed circuit boards 230
18.2.2.1 Board size 230
18.2.2.2 Functional division 230
18.2.2.3 Allocation of gates to packages 235

19 VERIFICATION OF LOGIC DESIGN 236

19.1 Interface specifications 236
19.2 Circuit checking 237
19.2.1 Numbering of drawings 237
19.2.2 Interconnection schedule 238
19.2.3 Logic checking 240
19.2.4 Variable data inputs 245
19.3 Errors in logic 245
19.3.1 Logical errors 245
19.3.2 Timing errors 246

20 APPENDIX 1 GUIDELINES FOR THE BEST USE OF T.T.L. 248
APPENDIX 2 ABREVIATIONS USED 251

RECOMMENDED FURTHER READING 253

INDEX 254

1

Introduction

1.1 T.T.L.—What it is

T.T.L. (Transistor–Transistor Logic) is the world's largest range of digital integrated circuits (or rather microcircuits). The range includes many different families of T.T.L., and each family provides a range of logic gate functions and flip-flops. Many of the families also include more complex functions which are known as M.S.I. (Medium-Scale Integration) devices.

The basic T.T.L. gate is a multi-input switching circuit, in which the electrical (or logical) state of the output depends on the logical combination of the electrical states of the inputs. All the families of T.T.L. are electrically compatible and units built from one version can be connected directly to units built from other T.T.L. families provided that fan-out rules are observed.

All devices in the T.T.L. range work from a single +5 volt power supply rail, and the two possible output states for any device are 'low', ON, or '0 level' at about a quarter of a volt above earth, and 'high', OFF, or '1 level' at 3–3½ V above earth.

T.T.L. is fast. The change in state of the output occurs in from 2 to 8 nanoseconds, about 6 to 20 ns after the input change which caused the output to change state. The timings depend on which family of T.T.L. is used, and on the load into which the device is driving. The input aperture, or the input voltages between which the output will be at neither the '0' or the '1' level, is narrow; the output impedance of T.T.L. is low (less than 100 ohms), and T.T.L. systems have excellent noise immunity. At around 12 milliwatts per gate, the power consumption of T.T.L. is moderate.

Most T.T.L. devices have an 'active pull-up' circuit, or 'totem-pole output stage'. This precludes the use of the 'wired-OR' (an OR function achieved by wiring together the outputs of two or more devices such that when any of the devices turns ON, the node adopts the ON condition regardless of the state of the other devices which feed the node), but it gives T.T.L. circuits the ability to drive into (moderate) capacitive loads without any significant loss of speed.

The inability to implement a 'wired-OR' is adequately compensated by the inclusion in all T.T.L. families of gates which have internal 'OR' functions. There are a few T.T.L. devices which have 'open collector' outputs which can be 'ORed' together (but which require an external 'pull-up' resistor) and there are also a few devices which have a tri-state output. These tri-state outputs have an extra transistor and diode which effectively inhibit the output stage and leave the output terminal at high impedance until the appropriate control signal is applied to enable the output. These devices can be 'wired-ORed' together, but the controls must be so arranged that only one output can be enabled. Open-collector and tri-state outputs are not considered in this book—only the normal 'totem-pole' output is described in detail.

T.T.L. has a good 'fan-out'. The number of T.T.L. devices which a gate can drive varies from family to family, but the minimum fan-out within any family is five (in families which have low and high fan-out gates).

Each T.T.L. device is a small chip (or die) of silicon, from about one to two and a half millimetres square, by about a tenth of a millimetre thick. These minute chips are mounted in larger cases which protect the device from environmental contamination and damage, and also simplify the job of mounting the devices on printed circuit boards. The most commonly used case is the 14-pin Dual-in-line Package (D.I.P.), which has its leads 0·100 in apart, in two rows 0·300 in apart, and which is mounted with its leads passed through holes in a printed circuit board. The alternative package form is the Flat Pack, a smaller body than the D.I.P., with flat ribbon leads in two rows on 0·050 in centres. Flat Packs are mounted by welding or reflow soldering their leads onto pad areas on the surface of a printed circuit board, or they can be stacked in 'cordwood' modules. Some flip-flops and complex elements are supplied in 16-pin versions of the D.I.P. or the Flat Pack, and some more complex M.S.I. functions are supplied in 24-pin D.I.P.s.

T.T.L. chips are also used in 'hybrid' packages, where a number of chips are mounted in a single package, together with the necessary internal interconnections to achieve 'custom designed' complex M.S.I. or L.S.I. (Large Scale Integration) functions.

T.T.L. devices are generally available with at least two working temperature ranges; −0°C to 70°C for 'commercial' or 'industrial' devices and −55°C to +125°C for 'military' devices.

T.T.L. is cheap. Basic gates can be bought for less than the cost of the discrete components which would be needed to make up the gate (if T.T.L. could be made from discrete components on a production basis).

Not only are the gates cheap to buy, they are cheap to use. Highly sophisticated engineering forms are unnecessary, and the small number of devices which have to be assembled to implement a system (compared with a discrete component system) reduces the design and assembly costs and increases the reliability. Some of the complex M.S.I. functions may seem expensive (on a per package basis), but when the costs of the basic gates which they replace are calculated and the cost of the board area saved is considered, it will be found that the more complex devices are, in fact, great money savers.

In the integrated circuit market, price is a function of demand (as well as supply). Overheads are heavy, design costs are high, and it is only when large volume sales can be made that device prices can be low. As more and more users realized the advantages of T.T.L., and more manufacturers entered the field, so the volume of production increased and prices fell. The current recession in the industry has led to what may well prove to be an all-time low in prices, but even without this recession T.T.L. devices would still have been amazingly cheap.

At the moment there does not appear to be any likely successor to T.T.L. Whatever device does supersede T.T.L. must offer very much better all-round performance, or very much cheaper manufacturing capabilities, because while T.T.L. prices remain low and there is competition between manufacturers the users of integrated circuits will be understandably loth to change to a new type of device for which the price must almost inevitably be higher, even if the price could ultimately undercut that of T.T.L.

1.2 Brief History

Microcircuits are a product of the 1960's. As semiconductor manufacturers mastered the intricacies of the planar silicon diffusion processes, they were able to put several transistors on the same chip, together with the necessary resistors and interconnections, and the integrated circuit era dawned.

Initially, the semiconductor manufacturers integrated the logic circuits which had previously been built up from discrete components, and D.T.L. (Diode Transistor Logic) and E.C.L. (Emitter-Coupled Logic) were the first families to establish themselves. However, it soon became apparent that the new production techniques could be used to implement circuits which could not be built from discrete components. In integrated circuits, it is not necessary to minimize the number of transistors used to achieve the most economical result. Transistors can be diffused at the same cost as diodes. Also, in integrated circuits the values of components might change with age (as do the values of discrete components), but all components on a chip would age and drift together, and so circuits could be designed to take advantage of the fact that parameters such as the ratio between the values of two resistors would remain practically constant throughout the life of the device.

T.T.L. was the first circuit to be designed specially for production as an integrated circuit. Sylvania's S.U.H.L. family was the first out, followed by Transitron's H.L.T.T.L., then Texas Instruments brought out their Series 54/74. The last family to appear was S.G.S.–Fairchild's 9000 range. Texas Instruments' marketing policy established the 54/74 Series as the major T.T.L. family, so that by the end of the 1960's most major semiconductor manufacturers were making Series 54/74 or a directly compatible family.

1.3 Alternatives to T.T.L.

The alternative forms of logic circuits to T.T.L. are (in descending order of speed), E.C.L., D.T.L., and M.O.S. devices. E.C.L. devices offer higher speeds than T.T.L., but this is at the expense of considerably higher power dissipation, and the faster E.C.L. devices usually need expensive, sophisticated mounting techniques if their speed potentials are to be realized fully. Also E.C.L. devices are considerably dearer than T.T.L.

D.T.L. devices are usually slower than T.T.L., consume slightly less power, and are dearer than T.T.L. The biggest disadvantage of D.T.L. is that accurate speed calculations are very difficult because the turn-off delay is very sensitive to capacitance on the output node. The two major families of D.T.L. available are compatible with T.T.L., and can be used in T.T.L. circuits when delays between about 30 ns and half a microsecond are required.

M.O.S. devices are generally an order of magnitude slower than T.T.L., but have the advantages of extremely low power dissipation and suitability for large scale integration. A number of manufacturers of M.O.S. complex functions provide their products with T.T.L. compatible input and output circuits.

When total system costs are considered, T.T.L. gives the best overall speed/cost compromise. Sophisticated engineering forms are unnecessary, and the low cost of T.T.L. packages makes parallel working economically practical. In fact in many cases faster working of a whole system can be achieved with T.T.L. than could be achieved at a comparable cost with the fastest E.C.L. available.

1.4 Manufacturers' Literature

All manufacturers of T.T.L. provide data sheets which describe most of the parameters of the devices in clear terms. However, there are notable omissions from some data sheets, of which edge speeds and supply rail current are the two which will most affect the average user of T.T.L. Many manufacturers offer 'typical' figures—but before designing any equipment it is advisable to check how typical these figures are! It has been the author's experience that when figures are not quoted on data sheets, the manufacturers are prepared to discuss these figures, and to accept a user's purchasing specification which rectifies the omission. If extra testing is involved, the price will usually be increased, but often the parameter involved may be accepted as not calling for extra testing.

Many manufacturers also supply 'Application Handbooks' or 'Manuals'. These should be read with caution, and it should be remembered that they are basically sales literature. In order to impress a potential customer, the authors of these application handbooks may sometimes present a somewhat one-sided picture of a competitor's product. In one case, a point is made in a manner which makes it quite clear to the reader why the product of the firm who publish the handbook is to be preferred. However, a closer inspection reveals that the point is valid only if the user constructs a special circuit to produce the effect which is 'cured' by the manufacturer. In fact the diagram concerned does not show a normal T.T.L. interconnection at all, although the accompanying text suggests that it does.

1.5 Use and Misuse of T.T.L.

T.T.L. is easy to use—for an engineer who has worked with high-speed discrete component circuits. Alas, it is very easy to misuse T.T.L. if the basic principles of its operation are not properly understood. Interconnections should be properly designed to handle the high edge speeds of T.T.L., and earth return paths and power supply lines also need to be properly designed. Printed circuit boards for T.T.L. devices need not be expensive, but if reliable equipments are to be built, the entire system must be designed to suit the characteristics of T.T.L. If the logic system is intended to work in a 'noisy' environment, such as in the vicinity of large electric motors, special screening precautions will probably be necessary. The advertised high noise-immunity of T.T.L. does not mean total rejection of all possible noise!

Discrete component circuits can be used to drive, or to be driven from, T.T.L. devices. Whenever this is done it is essential to appreciate what the T.T.L. circuit is, how it works, and what stray capacitances, etc. are associated with a T.T.L. device.

Excellent reliability can be achieved from T.T.L. systems, but this is possible only if care is taken in the selection of the devices to be used, and in the thermal and mechanical design of the system.

2

Manufacture, Testing and Reliability

2.1 Manufacture of T.T.L.

T.T.L. devices are manufactured by a planar epitaxial process. In this process transistor junctions are formed by successive diffusions of p- and n-type impurities into the face of an epitaxial layer which is grown on the surface of a silicon crystal. The diffusion areas are defined by windows etched in an oxide layer on the face of the crystal, which prevents the impurity from diffusing into the whole face of the crystal. After each diffusion, the oxide layer is re-formed, covered with photo-resist, and fresh windows are etched for the next diffusion. The result is that after diffusion the chip is covered by an oxide layer, the thickness of which varies according to the number of successive diffusions which have been made in any area. The steps in the oxide layer can be seen under a 100 times microscope, and the circuit on any chip can easily be traced. Resistors are formed by p-type (base) diffusion into the n-type epitaxial layer. In order to isolate the resistors, the n-type area which contains the resistors is connected to the 5 V supply rail, so that all the resistors form reverse-biased diodes to the 5 V rail. The individual circuit elements are electrically isolated from one another because all elements form diodes to the substrate or to the supply rail, and in normal operation these diodes are reverse-biased. (See Section 3.1.4.)

After diffusion, holes are opened through the oxide layer to contact all emitters, bases, collectors, resistors, and other elements of the gate. A layer of aluminium is then deposited over the oxide layer, and, after masking with photo-resist, the aluminium is etched away to leave a pattern of metal tracks to interconnect the elements of the gate. (See Fig. 2.1.)

All tracks from input and output terminals of the gate run to square or rectangular bonding pads, which are usually placed round the edge of the chip. (See Figs. 13.14 and 13.15.)

After metallization and the etching of the tracks in the metal, some manufacturers protect the face of the chip with a layer of silicon nitride or Silane.

Readers who wish to learn more about the fabrication of integrated

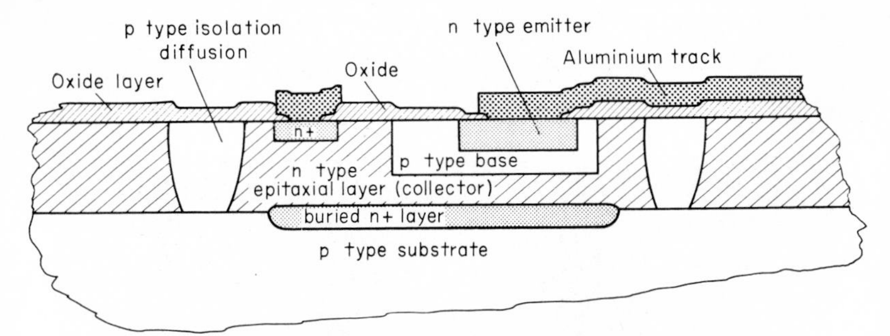

Fig. 2.1 Section through diffused transistor (not to scale).

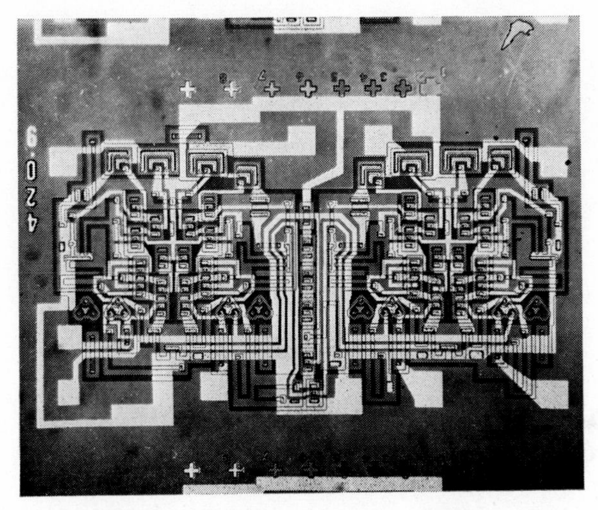

Fig. 2.2 Surface view of T.T.L. chip. (Courtesy of Sylvania Ltd.) Note that this chip has several transistors and resistors which are not connected by the metallization. An alternative metallization pattern can be applied to this chip to produce a different logic function.

circuits are advised to read *Analysis and Design of Integrated Circuits* and *Integrated Circuits—Design Principles and Fabrication*, both published by McGraw-Hill in the Motorola Series in Solid-State Electronics.

2.2 Packaging of T.T.L.

2.2.1 'FACE-UP' PACKAGING

2.2.1.1 *Separation of chips*

T.T.L. devices are not made one at a time. Over a hundred devices are laid out in a regular array on a wafer of silicon, and made simultaneously.

When all the diffusion processes have been completed, the finished wafer is scribed with a diamond point and broken into separate chips or dice. The chips are then mounted in the packages.

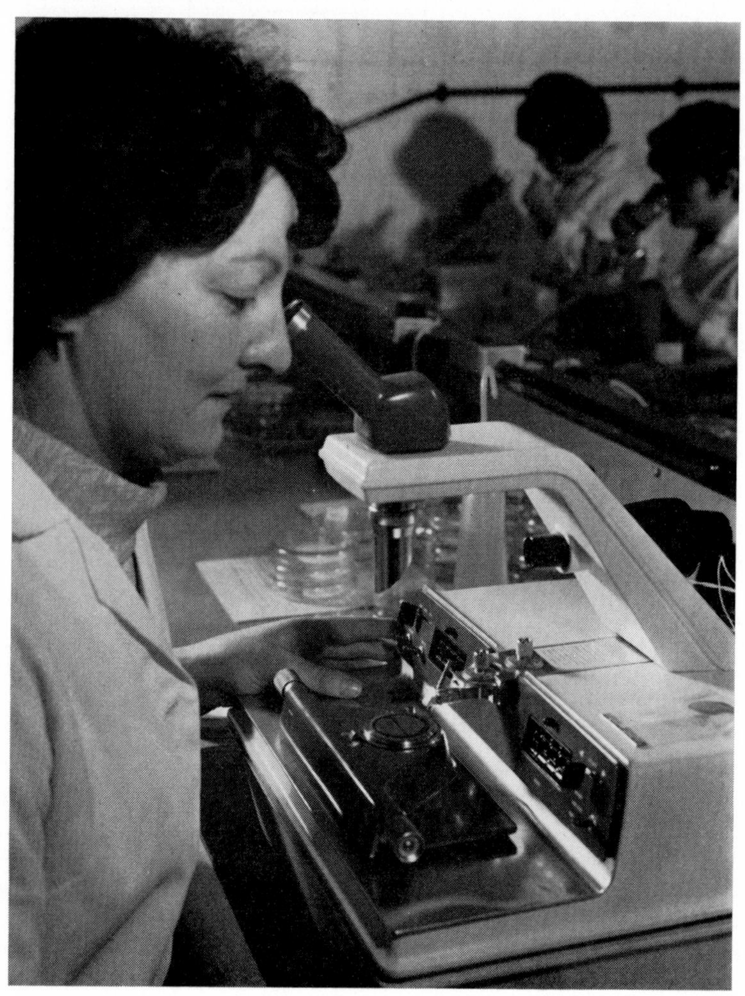

Fig. 2.3 Scribing a slice. (Courtesy of G.E.C. Semiconductors Ltd.)

2.2.1.2 *Packages*

The physical dimensions of the 14-pin Dual-in-line or Flat Packs normally used for the gates are described in *Printed Circuit Boards for Microelectronics* (by J. A. Scarlett, published by Van Nostrand Reinhold Company). Some M.S.I. devices are packaged in 16-pin versions of the same packages, with the same body dimensions, but the more complex elements are usually packaged in 'giant' D.I.P.s with 24 or more leads on 0·6 in centres instead of the 0·3 in of the 14- and 16-pin packages. Larger versions of Flat Packs are also used.

The packages may be ceramic, or they may comprise a plastic body moulded round the chip.

2.2.1.3 *Mounting and bonding*

The chip is first secured either to the base of a ceramic package, or to the lead frame for a plastic encapsulation. This is usually done by 'scrubbing' the chip down on to the gold plated base at an elevated temperature so the gold alloys with the silicon chip to give a good thermal contact. Electrical connections from the chip to the lead frame are made by wires, usually about one thousandth of an inch diameter. These wires can be aluminium, in which case they are ultrasonically bonded to the pads on the chip and to the lead frame, or they can be gold, in which case they are usually thermocompression bonded (ball bonded) to the chip and lead frame.

2.2.1.4 *Sealing and encapsulating*

After bonding, the lid of a ceramic package is sealed in place. If the device is to be plastic encapsulated, the lead frame with the bonded chip is placed in a transfer moulding press and the bodies are moulded.

2.2.2 FACE-DOWN PACKAGING

2.2.2.1 *'Flip-chip' bonding*

The chips can be mounted face downwards and connected to a substrate by means of small pillars, either on the chips or on the substrate, which connect the terminal pads on the chip to metallized tracks on the substrate.

The most commonly used substrate is a small piece of ceramic, on which metallized tracks are run from the package mounting pads along the edges of the substrate to the chip-mounting area. The chip can be hermetically sealed by a small ceramic 'pill box' which covers the chip and the ends of the metal tracks.

The pillars used in flip-chip bonding can be aluminium, or they can be silver and tin/lead. Some manufacturers of hybrid M.S.I. devices use flip-chip mounting for the active devices, but so far it does not seem likely that any device manufacturer will use the process to mount T.T.L. devices in their standard packages.

2.2.2.2 *'Beam lead' bonding*

Beam leads are little 'bars' or 'beams' of gold which are formed on the chip in such a way that they project beyond the sides of the chip. The chip is laid face down on a suitably metallized substrate, and the gold beams are welded to the ends of tracks on the substrate by special bonding machines.

The gold beams are formed after the last diffusion stage, before the wafer is broken into chips, and because of the space needed round each chip in which to form the beams, the chips must be more widely separated on the wafer than they are for 'face-up' or 'flip-chip' bonding.

The wafer can not be scribed once the beams have been formed, so the chips are separated by etching through the silicon from the back of the wafer, or by cutting with a laser beam. This eliminates all reliability hazards

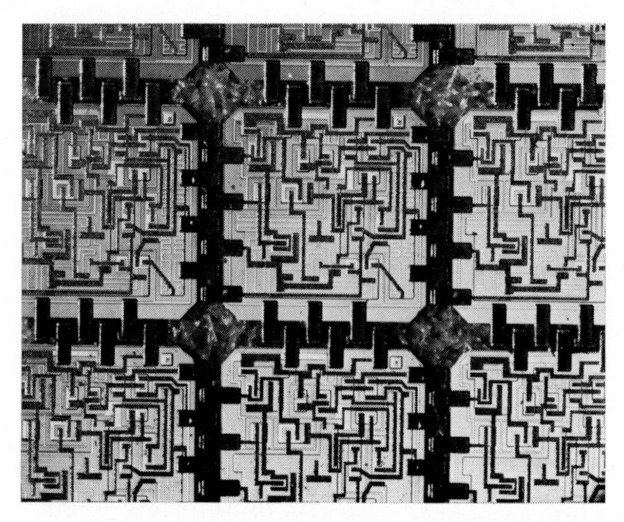

Fig. 2.4 Beam lead chips after back-etch. (Courtesy of Raytheon Company.) The chips are held to a gypsum matrix for wafer testing.

from cracked chips, and, because all beams are bonded to the substrate simultaneously, it is claimed that the reliability of beam lead bonded devices is better than that of conventional chips bonded face-up. The substrate on which the chip is mounted is similar to that for a flip-chip device.

One major United Kingdom supplier of T.T.L. has stated that a full range of T.T.L. gates and flip-flops will be available with beam leads in the near future (i.e. by early 1971).

2.2.3 HYBRID L.S.I.

Hybrid L.S.I. devices have several active chips bonded to a common metallized substrate. The chips can be bonded by any of the means described in Sections 2.2.1 and 2.2.2. The usual substrate is a piece of ceramic with pins along its edges so that it can be mounted on a printed circuit board in the same way as a D.I.P., or with gold contacts to fit an edge-connector along one edge. The metallization on the substrate will usually be multi-layer. At the time of writing, custom designed hybrid L.S.I. devices are available, but little is known about the long-term thermal and mechanical reliability of these large packages.

Further, more detailed, information on packaging can be found in the Proceedings of the Inter-Nepcon Conferences, which have been held annually at Brighton since 1968.

2.3 Testing of T.T.L.

2.3.1 PROBE TESTING

The devices are first tested when the slice is completed, before it is scribed and broken into chips. This testing is done on an automatic machine which is fitted with micro-probes. The slice is placed on the table on the machine, the probes are located over the first chip, brought down to contact the pads on the chips, and the device is tested. This test usually covers d.c. parameters and possibly a functional test. Speed is not generally measured at this stage. When one chip has been tested, the tester automatically steps on to the next chip, until the whole slice has been tested. Defective chips are marked with a blob of ink. Some firms use magnetic ink so the rejects can be removed automatically when the slice is broken.

2.3.2 CHARACTERIZATION AND Q.A. TESTS

The next stage of testing is usually a full d.c. parametric test on the completed packages. This is done on complex computer-driven testers, and it is at this stage that the devices are characterized as 'military' or 'commercial' grade devices. Complex elements, flip-flops, etc. are normally functionally tested. Speed may be measured only on a sampling basis unless a particular specification calls for 100 per cent testing. Other tests such as the ability to withstand shock, vibration, acceleration, and leak tests on hermetically sealed packages are also carried out on a sample basis.

Packages are not usually labelled until they are to be delivered because a batch may yield a high percentage of 'military' quality devices, and if most of the orders are for 'commercial' devices, the 'military' devices may be given the lower grade type number to meet the orders.

Full 100 per cent testing is normally carried out at room temperature (25°C) only, and appropriate allowances are made for parametric degradations which will occur at the rated temperature limits. Most firms test samples from all batches at the limit temperature as part of their quality control. However, when devices are to be brought to work at or near the upper limit of their temperature range, the test procedure adopted should be checked very carefully. One widely used technique is to store the

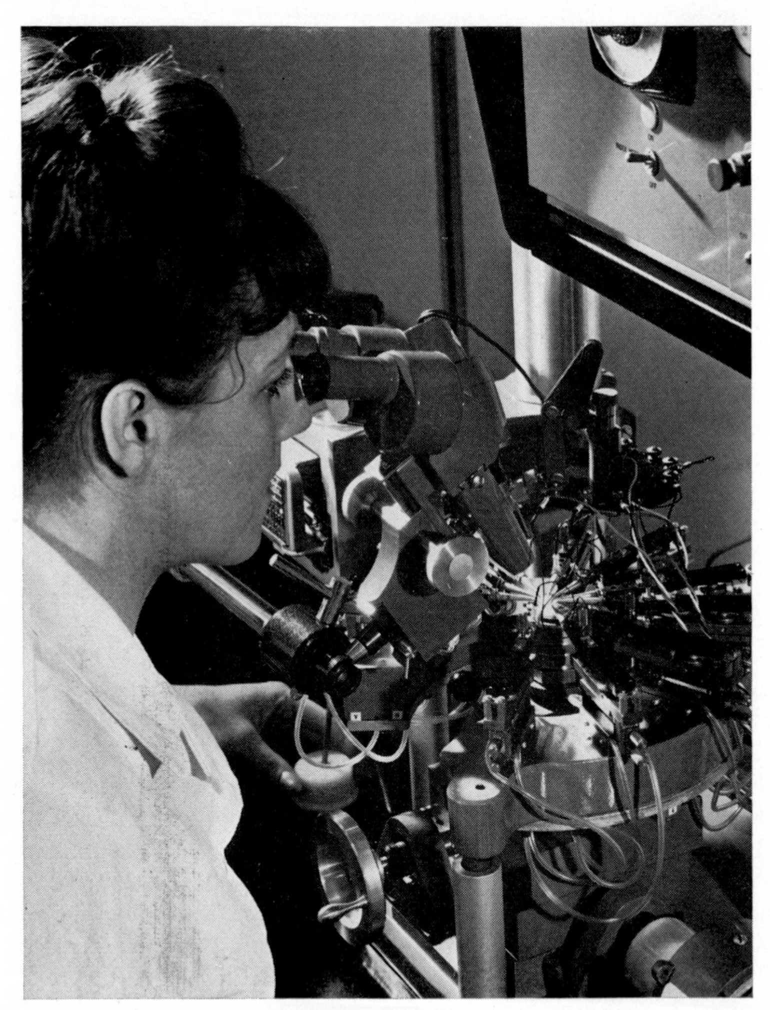

Fig. 2.5 Probe testing a completed slice. (Courtesy of G.E.C. Semiconductors Ltd).

devices at the limit temperature for sufficient time to be quite certain that the entire package and chip are at the rated temperature, then the devices are tested on a high-speed tester at this temperature. The test is carried out very quickly, and the package does not have time to warm up as it would if it was run continuously at the rated temperature. So far as the author is aware, an allowance is usually made on the test figures which should guarantee that the device will be within specified limits when it is run in an ambient temperature at the rated limit. On M.S.I. devices the chip could then be 30°C or more above the temperature at which it is actually tested.

Since test procedures and allowances for parametric variations vary from supplier to supplier a would-be customer should check on what tests are done. All firms visited by the author run very good quality control

Fig. 2.6 Probe tester heads and wafer under test as seen under low power magnification. (Courtesy of G.E.C. Semiconductors Ltd.)

programmes, but before devices are ordered for use in critical environments it is advisable to check that the supplier's Q.C. covers all the purchaser's requirements—assumptions on quality and reliability can turn out to be expensive guesses!

The tests of d.c. parameters do not necessarily determine the actual value of a given parameter, but they check that the parameter is within its rated limit. For instance, input threshold voltages are checked by holding the input at the specified voltage (with any guard band allowances) and measuring the output voltage. This test would usually be done with the device sinking or sourcing the full rated output current, so that the single

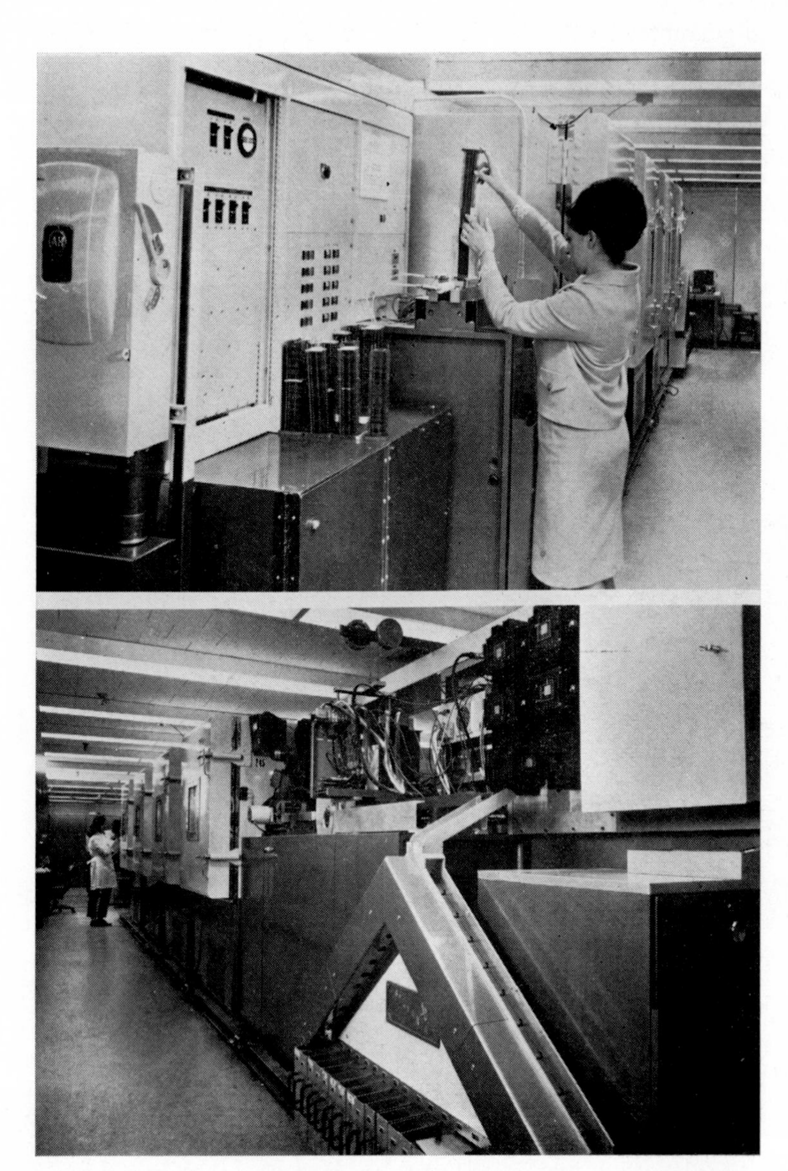

Fig. 2.7 'Mr. Atomic'—Sylvania's automatic tester for integrated circuits. (Courtesy of Sylvania Ltd.) Note circuits stacked ready for loading into the tester; four controlled temperature chambers, and sorting mechanism for tested circuits at output end of tester.

test verifies at least three major parameters. All such tests are 'GO-NOT-GO' tests, and even if the tester is coupled to a recording instrument, they do not show by how much the device passes the tests. Qualitative measurements to determine such points as the threshold at which a device actually starts to switch, or the load current an output can carry safely, must be done on specially built testers (which need be no more sophisticated than controllable voltage and current sources, a socket for the device to be tested, and a multi-meter).

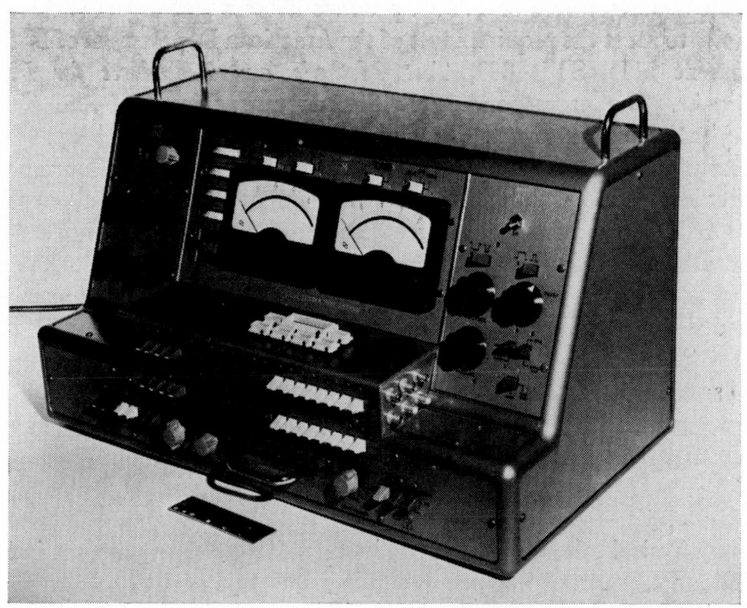

Fig. 2.8 Integrated Circuit evaluation tester designed by the author. (Courtesy of the Marconi Co. Ltd.) This tester can be used to evaluate expander packages together with the gates they expand, and can be used to investigate all d.c. and a.c. parameters. It can be programmed to test devices in D.I.P.s with up to 40 pins. All tests are selected and controlled manually.

Qualitative speed tests can be carried out on equally simple test rigs, such as a ring of gates, connected so that there will be an odd number of inverting gates when the device under test is inserted in the test socket, and a high speed (sampling) dual-trace oscilloscope. On such astable ring testers, the device under test will normally be loaded with a fan-out of one gate input and the oscilloscope probe. Higher loads can be added as desired, and the effects of leaving unrequired inputs open-circuit, etc. can be measured very easily. Variations of parameters with switching frequency can be measured on a three-inverting-gate ring in which two variable delays have been inserted.

2.4 Quality of T.T.L.

The manufacture of an integrated circuit which will work at all demands a high degree of quality control. Cleanliness and accuracy are essential at all stages, and any carelessness can result in a defective device. The details of mask design and manufacture vary from one manufacturer to another, and there are differences in the quality, reliability and consistency of their products.

Most manufacturers of T.T.L. produce devices and organize their test facilities to meet the requirements of the American Department of Defense document MIL–STD–883, *Test Methods and Procedures for Microelectronics*. The corresponding British document is BS 9400, *Generic Specification for Integrated Electronic Circuits of Assessed Quality*. Manufacturers' data sheets do not normally carry any guarantee of any specific quality level, and unless a quality level is defined, quotation of clauses from MIL–STD–883 will not guarantee quality—the standard defines only the test methods and inspection procedures to be used. It is left to the user to define quality levels in the purchasing specification or document. It is intended that detail specifications under the BS 9400 scheme will define quality levels, and purchasers of devices which carry a number in the BS 9400 series will be assured that certain quality tests have been carried out without any need for a purchasing specification.

2.4.1 MANUFACTURING DEFECTS

The physical tolerances involved in manufacture are very tight indeed; diffusion faults can and do occur, and in any crystal of silicon there may be faults in the lattice structure which can cause devices to fail. Faults can also be caused by minute irregularities in the photo-resist which is used to mask the slice while the oxide layer is etched, and any such faults are liable to cause early failures if they do not cause a chip to be rejected at the first probe test. Many of these faults are weeded out when the slice is probed, but the probe test may not fully test a more complex device, and defective chips can get through.

A further hazard arises when the slice is broken into chips. Chips from some manufacturers usually have clean, straight, edges, but others are more ragged, and there is a risk that fine cracks can propagate into the working area of the chip. Cases have been observed where the only visible fault on a device which has failed in service has been a very fine crack at one edge.

Bonding can also introduce faults. Bonding is done by operators who use powerful binocular microscopes and micromanipulators, but it is nevertheless a manually controlled operation and as such suffers from an inevitable human error factor. Bond wires can be run to wrong pins in the package, or the bonds may not be perfectly formed.

2.4.2 BURN-IN

All these faults can cause 'dead on arrival' devices, or devices which 'die' very early in their working life. Most of the real risk of any 'early deaths' can be avoided by 'burning-in' the devices; i.e. by running them at an elevated temperature for about 160 hours, after which the full characterization testing is done. So far as is known by the author all suppliers of T.T.L. devices offer 'burn-in'—at an appropriate extra cost.

Some users insist on burn-in, and claim that without it the costs of replacing defective devices at the board-test stage would be prohibitive. Other users claim exactly the reverse—that they find it cheaper to buy 'standard' devices at the minimum price possible, run the completed boards for a time before testing them, then replace the 'dead on arrivals' and the burn-in failures. Obviously the decision on whether or not to pay extra for burn-in depends on many factors, and the economics of 'burn-in' must be determined according to the circumstances involved. One point which must be remembered is that burn-in will not guarantee 100 per cent good devices, and faulty devices will still be found when the boards are tested. Even if the yield from the device manufacturer was 100 per cent, some devices might get damaged during assembly of the boards.

2.4.3 'GOODS-INWARDS' TESTING

Some users of T.T.L. claim that they have to do 'goods-inwards' testing on all devices bought, and 'dead-on-arrival' rates of up to 23 per cent have been quoted. Other users say that such testing is quite unnecessary, and quote figures such as '3 dead in over 800' and '1 bond failure in 600 packages'. Some users have their own tightly defined purchasing specifications, and rely on the supplier's Q.C. staff to ensure that these specifications are met. It has been the author's experience that suppliers of T.T.L. are prepared to allow a customer's accredited inspector to check on the in-house Q.C. program. The quality, price, and delivery time of T.T.L. vary from one manufacturer to another, and users must decide for themselves whether it will be better to accept the cheapest devices with possibly a poor yield after full 'goods-inwards' testing, or whether to pay more initially to get a higher yield without the expense of 'goods-inwards' testing.

One point concerning 'goods-inwards' testing which needs careful consideration is the question of marginal parametric failures. All testers used for 'goods-inwards' testing must be carefully callibrated and checked against the testers used by the device suppliers. If this is not done, it is possible that devices which are actually marginally inside their specified limits might be rejected. The author has found that all T.T.L. suppliers who have been asked have been willing to supply fully characterized callibration samples of gates and flip-flops.

Even when devices are marginally out of specification on one or more parameters, it may be that these devices would work quite adequately when they are installed in a system. Thus a user who does his first testing on fully assembled boards might not be aware that some of the devices were 'out of Spec.', and he would find a lower defect rate than if he did full 'goods-inwards' testing. The inclusion in an equipment of devices which are marginally 'out of Spec.' can result in a reduction in the long-term reliability of the equipment.

2.5 Reliability of T.T.L.

Accelerated life tests show that T.T.L. devices are very reliable, but their real long-term reliability in service is still undetermined. Engineers talk of 'ten year reliability', but although all the indications are that this (and much more!) will be achieved, a few more years must elapse before it is fully proven!

2.5.1 MOISTURE AND PLASTIC PACKAGES

Some reliability hazards are well recognized. Moisture or ionic contamination (especially sodium or chlorine ions) on the face of a chip can cause failure, and several manufacturers of high reliability equipment have banned plastic encapsulated devices because it is now well established that the usual transfer-moulded encapsulation does not protect the chip from the ingress of moisture, and the plastic can yield free ions which may accelerate failure.

Plastic encapsulations are also regarded as a reliability hazard because it is thought that the plastic can cause dangerous stresses in the bond wires. Some manufacturers of plastic encapsulated devices claim that this is untrue, but one leading supplier of T.T.L. in plastic D.I.P.s uses ball bonded gold wires because it was found that ultrasonically bonded aluminium wires would not withstand the moulding stresses.

2.5.2 'PURPLE PLAGUE'

The use of gold wires introduces another reliability hazard; the risk of gold-aluminium intermetallic compounds forming, at elevated temperatures where the wire is bonded to the chip. These intermetallic compounds, some of which also involve silicon, can cause a connection to go high impedance or even open circuit. Several papers have been written about these compounds, but the average user of T.T.L. need only be aware of the fact that if he allows chip temperatures to rise over about 150–170°C, intermetallic compounds ('purple plague' or 'black plague') may be formed if gold bond wires have been used. Aluminium bond wires do not cause a reliability hazard in this manner.

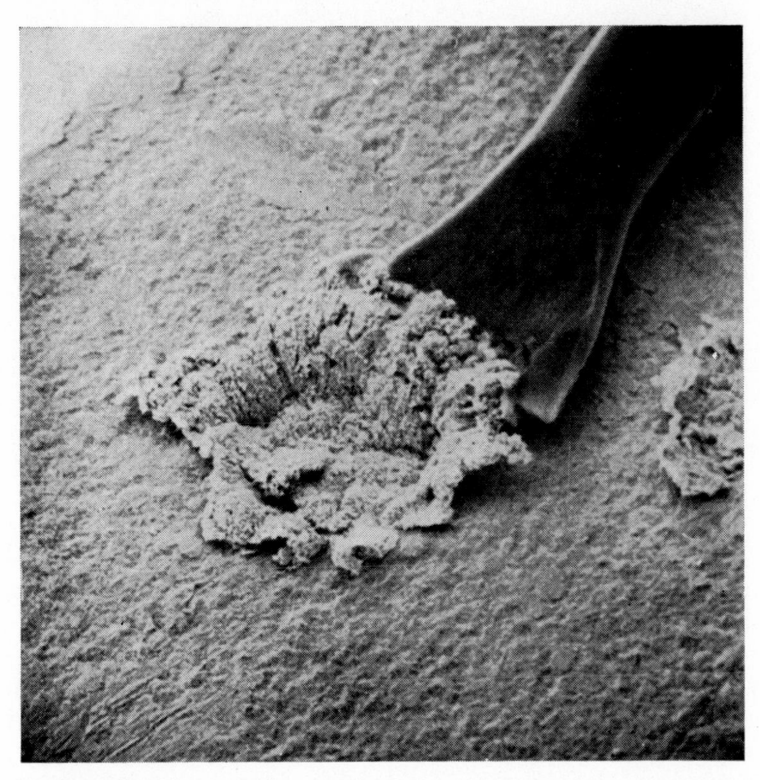

Fig. 2.9 Scanning electron micrograph of aluminium wire bonded to a gold post showing plague formation after heating at 250°C. (Courtesy of Marconi Co. Ltd.)

'Purple plague' is not a hazard when gold beam leads are used, because the gold beams are formed over layers of other metals (platinum and titanium) which isolate the gold from the face of the chip, and there is no aluminium present.

2.5.3 METAL MIGRATION

At elevated temperatures and when passing current, the aluminium which forms the tracks on the face of a chip can migrate until the width of

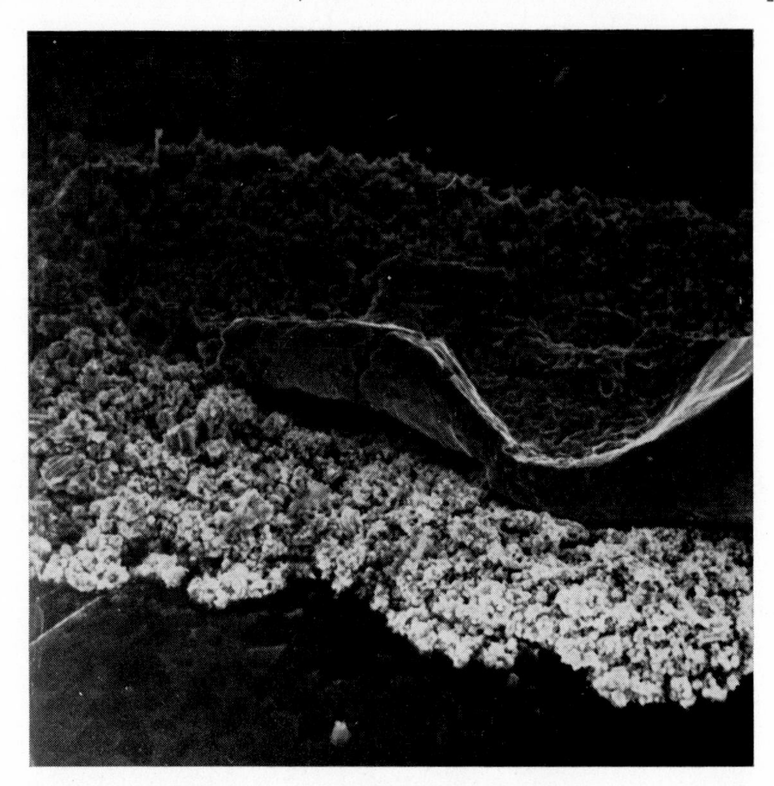

Fig. 2.10 Scanning electron micrograph of gold wire bonded to an aluminium contact pad, showing plague formation after heating at 250°C. (Courtesy of Marconi Co. Ltd.)

a track is reduced to the point at which it will fuse. This should not normally be a hazard on T.T.L. devices, but a chip may start life with a defect which reduces the width of a track, and if the track is passing a high current, the device can fail. This effect is accelerated at elevated temperatures.

If high reliability equipment is to be built, it is worth verifying that all tracks on the faces of the chips are amply wide for the currents they will have to carry. Figures of 60—75 mA per thousandth of an inch width have been suggested as a safe limit figure. This may sound a high current for devices with a normal input current of just over a milliampere, but it must be borne in mind that the earth return track of a fully loaded hex-inverter package will be carrying current from sixty such inputs!

Current densities in the bond wires should also be checked. On some M.S.I. devices the supply rail current density in the bond wires can be over 30,000 A/in^2, and the earth return on a quad lamp driver could run up to over 100,000 A/in^2 if circuit designers are not warned in advance of this reliability hazard. Bond wires are usually between 0·001 and 0·0015 in in diameter, so as a very rough guide, every two milliamps of current into or out of a terminal could mean a current density of 1000 A/in^2. All doubtful cases should be checked with the device manufacturers. They might be using double bond wires, or an earth return path via the chip bond to the lead-frame.

2.5.4 JUNCTION TEMPERATURES

Elevated temperatures (below 200–300°C) do not appear to have any adverse effect on the long-term life and reliability of a silicon junction provided that the device is free of all possible manufacturing defects and the junction is absolutely protected from all possible contamination by impurities. Such absolute perfection is hard (if not impossible) to achieve, and devices which are allowed to 'run hot' can have their lives shortened by accelerated diffusion of impurities into the junctions.

However, there is another point which must be considered before it can be assumed that chip temperatures need not be controlled. The intrinsic leakage current of a silicon transistor can be taken as doubling for every 10°C rise in temperature, from about 0·1 to 0·5 nA at 20°C. A rise in temperature of 100°C will cause a thousandfold increase in this leakage current, and a rise of between 210°C and 240°C will increase the leakage current to over 1 mA.

This increased leakage current might be sufficient to turn on the phase-splitter stage of a gate or M.S.I. chip, and could cause the device to fail to function (see Section 3.1.2). At about 220°C chip temperature, leakage currents may be such that noise margins in the device will be dangerously reduced.

It must be borne in mind that when the manufacturers of T.T.L. offer devices to work up to 125°C, the testing and reliability of operation of the device are based on the assumption that a single device in free air at 125°C is being considered. When devices are closely packed on printed circuit boards any one device on a board will be operating in an ambient temperature considerably higher than the temperature of the air entering the equipment. All effects of temperature discussed in Section 10.1 are based on the assumption that the effective ambient temperature is restricted to the normal operating range specified by the manufacturer.

Because there must always be some risk of device failure caused by accelerated contamination of the junctions, it is advisable to limit chip temperatures. Most manufacturers specify an absolute limit storage temperature of 150°C, and in high reliability equipments, this is a good limit figure to aim for. (Some authorities regard chip temperatures between 110 and 125°C as the highest that should be allowed if long-term reliability is required.) Some M.S.I. devices can dissipate over half a watt, and it might be impracticable to limit their junction temperatures to 150°C. Chip temperatures up to 200°C would appear to be acceptable provided gold bond-wires have not been used. Until sufficient time has elapsed for firm end-of-life data to be accumulated, chip temperatures above about 150°C should be regarded as a potential hazard to long-term reliability.

The best information available at the time of writing suggests that at 200°C the mean time to failure might be reduced to a quarter of the M.T.T.F. at 100°C, but all figures seen by the author have been based on 'no failures after so many thousand (or million) device hours', and until a substantial number of end-of-life failures have been recorded, any such figures must be regarded with suspicion.

2.5.5 THERMAL GRADIENTS

Calculations of chip temperatures are possible only when the chip to case thermal gradient is known. Many manufacturers quote a value of 0·1°C/mW for 14-pin Dual-in-line packages, and tests on 14- and 16-pin packages have indicated actual values in the region of 0·075 to 0·085°C/mW. Some manufacturers have quoted 0·3°C/mW junction to air temperature gradient. However, since the environment is not defined, this figure is of little value. The results of tests known to the author indicate that this might be a reasonable figure for packages mounted on boards, with the packages on about one inch centres and the boards on about half inch centres, with free air circulation. Single devices in free air appear to have chip to air temperature gradients of around 0·14°C/mW (i.e. about 0·06°C/mW case to air). If M.S.I. is to be used in any high-reliability equipment, the junction (or case) to air temperature gradients for the proposed engineering form must be established at the earliest possible stage in the design, because simple calculations will show that without a well-thought-out cooling system, the chip temperatures in some M.S.I. devices could rise to well over 200°C with the equipment in air at 50°C.

When chip temperatures are calculated, operating frequency and output load currents must be taken into account, as explained in Sections 9.5.1 and 9.5.2, and appropriate allowances must be made for heat conducted via the board from other devices and for the fact that the coolant will be warmed up as it flows through the equipment.

3

The Basic T.T.L. Circuit

3.1 Circuit Description and Logical Operation

The basic T.T.L. circuit is shown in Fig. 3.1. It comprises three main areas—the input 'AND' function associated with VT1, the phase-splitter or 'OR' function associated with VT2, and the output amplifiers D3, VT4 and VT5.

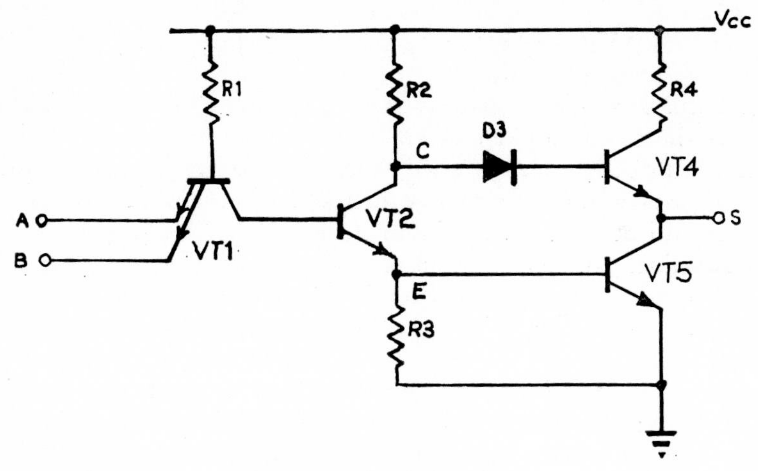

Fig. 3.1 Basic T.T.L. Gate

3.1.1 THE INPUT 'AND' FUNCTION

The multi-emitter transistor VT1 can have as many separate emitters as are required for the gate type being produced, the normal maximum number being eight. Two inputs only are shown on Fig. 3.1 as the two-input gate is the normally used 'building brick' for most logic circuits, and throughout this book it is the type most commonly used for descriptions. The action of the input stage is most easily understood if it is redrawn with the collector and emitters shown as diodes connected to the base, as in Fig. 3.2. Then it is clear that if both input diodes are returned to a high voltage (4–5V), current through R1 can flow through the collector diode to provide base drive for VT2. If either of the input diodes is connected to earth or to a low voltage point which is capable of sinking current the current through R1 will flow out through that diode; the base

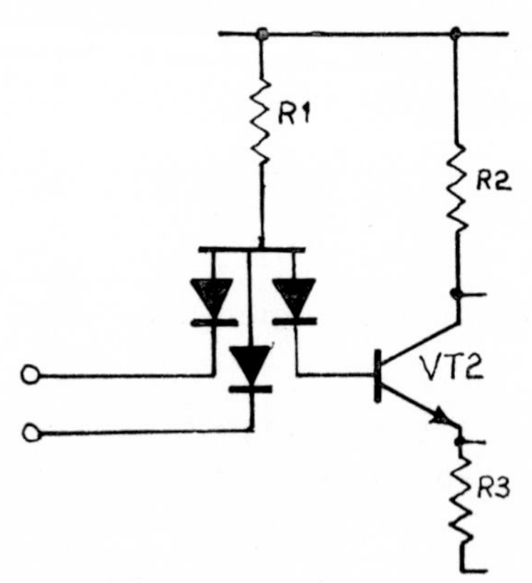

Fig. 3.2 Diode equivalent of input transistor.

of VT1 will be at a low voltage (V_{BE}VT1 above the input potential), and there can be no base drive to VT2.

Thus with both inputs at high voltages the collector of VT1 will be high, whereas if either of the inputs is taken to earth or a low voltage the collector will be low; i.e. the input transistor performs a positive logic 'AND' function.

3.1.2 THE PHASE-SPLITTER OR 'OR' FUNCTION

In the simple two-input gate shown in Fig. 3.1 the second transistor VT2 is a conventional phase-splitter stage. Resistors R2 and R3 are usually roughly equal in value, and are between a quarter and a third of the value of R1, so the transistor need not have a very high current gain to be able to saturate. When VT1 is supplying base drive to VT2 (i.e. when all inputs are returned to a high voltage) point E (the emitter of VT2) can rise only to V_{BE}VT5, and point C (the collector of VT2) is held down at V_{BE}VT5 + $V_{CE(sat)}$VT2. When VT1 cuts off the base drive to VT2, only leakage current can flow through R2 and R3 so points E and C can be at (nominally) earth and V_{CC}. An 'OR' function can be achieved by connecting two or more such phase-splitter transistors in parallel. Fig. 3.3 shows how this is done in a 'Triple Three' gate where the 'ANDs' of groups of three signals are 'ORed' together at points C and E. If any of the three input transistors VT1A, VT1B, or VT1C has all its three inputs returned to a high voltage, its corresponding phase-splitter transistor will be saturated and the other two phase-splitters will be cut off. To turn off the phase-splitter stage, one input to each of VT1A, VT1B, and VT1C must be earthed or returned to a low voltage. The number of functions which can be 'ORed' together is limited by the leakage current through each of the phase-splitter transistors. If too large an 'OR' function is attempted the total leakage current may cause point E to rise to a level where the output transistor is on the verge of conduction.

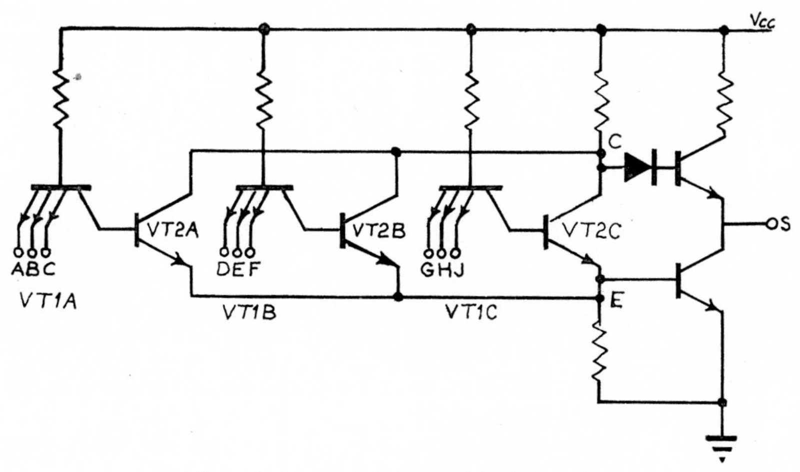

Fig. 3.3 Gate with internal 'OR' function.

3.1.3 THE OUTPUT STAGE

The main output transistor is VT5. When all the inputs to the gate are high, VT2 will be 'ON', and base drive will be supplied to VT5, which will also be 'ON'. As was explained in Section 3.1.2, point C will be at B_{BE}VT5 + $V_{CE(sat)}$VT2. If we disregard VT5 and consider only D3 and VT4, point S would be V_{D3} + V_{BE}VT4 below point C. V_{BE}VT4 will not be as large as V_{BE}VT5, but V_{D3} will be considerably more than $V_{CE(sat)}$VT2, so point S could be slightly below earth. Since we can not disregard VT5, the output potential of the gate will be $V_{CE(sat)}$VT5 when all inputs are high.

When any input is low, VT2 is 'OFF', and there is no base drive to VT5. VT2 is taking only leakage current through R2, so point C is at almost V_{CC}. If there is a small current flow to earth from the gate output, point S will be V_{D3} + V_{BE}VT4 below point C. Base drive to VT4 will be negligible, so the potential drop across R2 caused by this base current can be ignored and it can be said that when any input is held low the output of the gate will be two semiconductor drops below the supply rail voltage. There are several different pull-up circuits used in the various T.T.L. families, but all of them feature two cascaded semiconductors between V_{CC} and the output. The resistor R4 is included in the output circuit to limit the current which can flow if the output of an OFF gate is shorted to earth.

Because the basic T.T.L. gate requires all its inputs to be high for the output to be low, it performs the positive logic 'NAND' function, where

$$\overline{S} = A.B \qquad \text{or} \qquad S = \overline{A.B} = \overline{A} + \overline{B}$$

A gate with an internal 'OR' function as in Fig. 3.3, is best described as an 'AND-OR-INVERT' gate, which performs the function

$$\overline{S} = A.B.C + D.E.F + G.H.J \text{ or } S = \overline{A.B.C} + \overline{D.E.F} + \overline{G.H.J}$$

3.1.4 CAPACITY AND PARASITIC DIODES

The basic T.T.L. gate can be considered as the simple circuit shown in Fig. 3.1 for most practical purposes, but this is far from the whole story. Every terminal of the device exhibits capacity to the earth terminal and to V_{CC}, and through the entire circuit there are stray capacitances to earth and to other parts of the circuit. Since the whole circuit is diffused into a single silicon substrate each transistor collector is formed as a

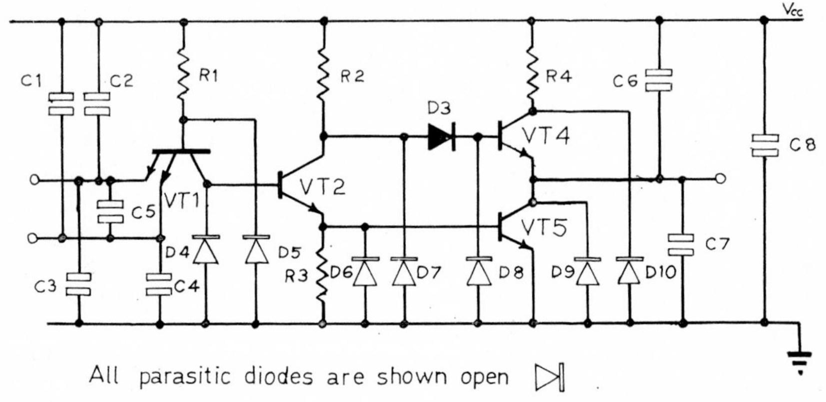

Fig. 3.4 Stray capacitance and parasitic diodes in T.T.L. gate.

reverse-biased diode to the substrate, and all diffused resistors are associated with a similar reverse-biased diode distributed along their length. The user of T.T.L. is not concerned with all the possible stray effects, but a knowledge of some of the more basic strays helps an understanding of the behaviour of a device. The main stray capacitance effects and parasitic diodes are shown in Fig. 3.4, where the diodes associated with resistors are shown lumped at one end. Where a resistor's parasitic diode would appear in parallel with that for a transistor collector, only one diode is shown.

The parasitic diodes which noticeably affect the terminal performance of the devices are D4 and D9, which limit the negative excursions of the inputs and outputs.

The stray capacitances shown are those which similarly affect the terminal performance of the device. The effective capacity between V_{CC} and earth, C8, has been measured on several hundred packages, and a value of about 64 pF per package found. With this capacity shunting the capacitors between an input and earth and the input and V_{CC} it is not possible to assign individual values to both the input capacitors. The combined effects of stray capacity at an input can be represented by a single capacitor to earth with a value of about 2·7 pF (mean of a number of measurements). It is probable that this is mainly capacity to earth through the input transistor. At the output, evidence of tests indicates that the two capacitors are more nearly equal in value, with C6 and C7 each having a value of about 3·5 pF, giving a total effective capacitance of 7 pF to earth. Capacity from input to input, C5, is minimized by the design of the input transistors.

3.2 Family Differences in the Basic Gate

This section is not intended to be an exhaustive catalogue of suppliers of T.T.L. Generally the manufacturers mentioned are those with whom the author has had some dealings in connection with T.T.L. The only significance attached to the order in which the families are presented is that it is roughly the order in which the author worked on or heard about the various circuits.

Where resistor values are quoted these have been taken from the data sheets or from information disclosed at symposia or sales presentations. In some cases the information available is conflicting, and there may be minor errors. Since the diffused resistors have a very wide tolerance, and several published circuits are at best approximations to the chip layouts, it is believed that any such errors which may be present will be of no practical significance. Because of the wide tolerance on resistor values, all current levels quoted in this section must be regarded as a typical guide only. On no account should the levels quoted here be used in any toleranced design calculations.

No attempt is made here to explain the significance or the working of variations in the circuits. These are dealt with in the appropriate sections of other chapters.

Some of the circuits drawn show 'overswing' diodes on the gate inputs (see Section 4.1). The presence (or absence) of these diodes on the drawing for a particular family of logic should not be taken as implying that they are included (or not included) by all manufacturers listed as making that type of logic. By the time this book is published it is possible that some manufacturer, who is not including these diodes at the time of writing, will have modified his masks to incorporate diodes.

Throughout this book, the same component numbers are used for similarly positioned components in the basic gate circuits. Thus VT2 is always a phase-splitting transistor, VT4 is always the output pull-up transistor, and so on. This means that on some circuits the component numbers will not run consecutively.

Most manufacturers offer 'Military' devices with a working temperature range of −55 to +125°C, and 'Industrial' or 'Commercial' devices with the working temperature range limited to from 0 to +70°C. In some cases the specifications are different for the two ranges, but the circuits are the same.

3.2.1 THE S.U.H.L. 1 GATE

S.U.H.L. 1 (Sylvania Universal High Level logic) is a 20 ns T.T.L. manufactured by Sylvania, a subsidiary of General Telephone and Electronics. S.U.H.L. 1 is available with a fan-out of 6 or 12 (see Section 9.3), and the power consumption is moderate (nominally 15 mW per gate).

The S.U.H.L. 1 gate is the circuit used to illustrate 'the basic T.T.L. gate' (Fig. 3.1). Nominal resistor values are shown in Fig. 3.5. There is some uncertainty in the value of R4, which might vary slightly from

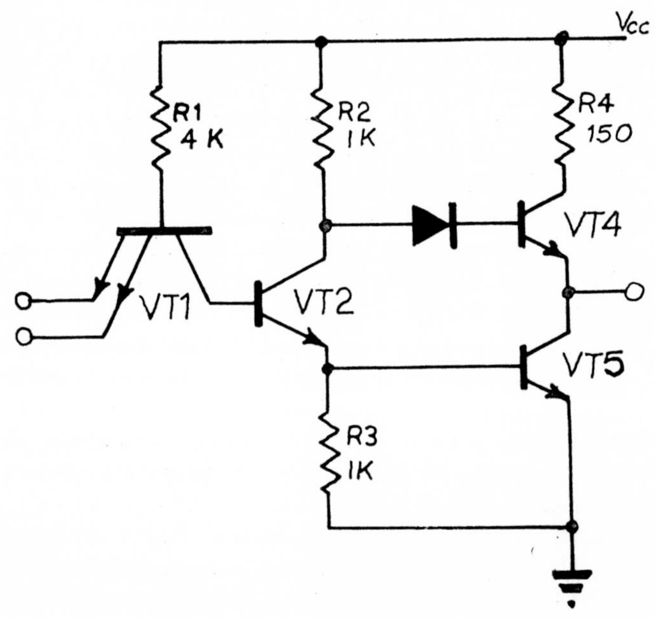

Fig. 3.5 Basic S.U.H.L. 1 gate.

manufacturer to manufacturer, between 125 Ω and the 150 Ω shown here. Similar devices following the same circuit diagram are M.T.T.L. 1 made by Motorola Semiconductor Products, RAY 1 made by Raytheon, and H.L.T.T.L. 1 made by Transitron Electronic. Philco-Ford and Westinghouse are also known to make nominally compatible devices, but the author has no personal knowledge of their products.

M.T.T.L. 1 and RAY 1 appear to be exact copies of S.U.H.L. 1 (in specification, range of devices and package pin allocation), but H.L.T.T.L. 1. has differences in its specifications and range of devices offered. H.L.T.T.L. and S.U.H.L. can be regarded as being fully compatible, the only effect of the differences in the specifications being a slight loss in fan-out capability if worst-case toleranced designs are used.

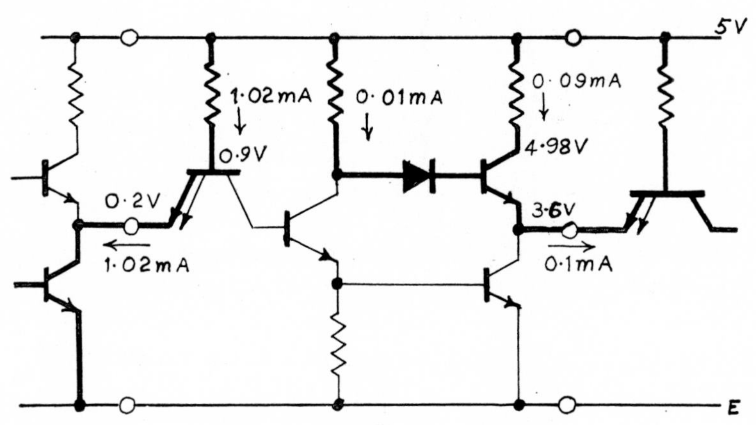

Fig. 3.6 Current flow in 'off' gate.

Higher speed versions of S.U.H.L. exist (see sub-section 3.2.4 *et seq.*), and these can be mixed with, or substituted for, S.U.H.L. 1 (provided that the printed circuit boards carrying the devices have been designed to avoid problems of cross-talk and ringing associated with the higher speed circuits).

Nominal current and voltage levels in a S.U.H.L. 1 gate in the OFF state are shown in Fig. 3.6. One input is shown connected to an ON gate whose output is at 0·2 V, and the output is shown driving one input of a further gate. The input leakage current of the latter gate is assumed to be at its maximum specified value. The supply rail is assumed to be at 5·0 V, and base-emitter drops of 0·7 V are assumed throughout. Normal transistor leakage currents are assumed to be negligible. The exact division of the output current between R2 and R4 depends on the h_{FE} of VT4, and the exact output voltage depends on the current flow through R2. The total current taken from the supply rail by a gate in the OFF state is nominally

1·12 mA, but this will increase as the current taken from the output increases. It will also increase slightly if the 'low' input is returned to earth instead of to 0·2 V.

ON state conditions with the same assumptions are shown in Fig. 3.7. This shows a nominal current consumption of 4·7 mA, and 4·1 mA base drive to VT5—clearly adequate to guarantee saturation of VT5 at the full specified fan-out. The current consumption is independent of the output load in the ON state.

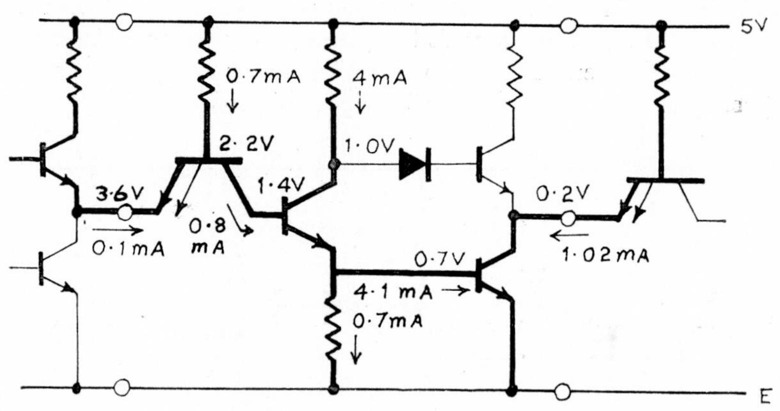

Fig. 3.7 Current flow in 'on' gate.

3.2.2 THE SERIES 54/74 GATE

Series 54/74 T.T.L. (See Fig. 3.8) was first introduced by Texas Instruments. It is slightly slower than S.U.H.L. 1, but its power consumption is slightly less (nominally 10 mW). The fan-out of all gates is 10. The main difference from the S.U.H.L. 1 circuit is that D3 is on the emitter of VT4 instead of in its base circuit, and the value of R2 has been raised to 1·6 K. Input leakage currents are specified as being less than for S.U.H.L. 1.

Current levels in R1 are the same as those in the S.U.H.L. 1 gate, but the lower specified input leakage current reduces the nominal OFF state current consumption to 1·06 mA. The higher value of R2 reduces the ON state supply rail current to 3·2 mA, but it also reduces the base drive to VT5 to 2·54 mA.

Fully compatible equivalents to Series 54/74 are made by Fairchild, Ferranti (Micronor 5), I.T.T. Semiconductors (MIC 54/7400 Series), Marconi-Elliott Microelectronics, Motorola, Mullard (FJ Range), National Semiconductors (DM 8000 Series), S.G.S., Sprague, Sylvania, and Transitron (H.L.T.T.L. 4). At the time of writing, Series 54/74 and its equivalents account for the biggest share of the world market for T.T.L.

Whereas Series 54/74 circuits work in a very similar manner to S.U.H.L., and the families can be electrically interconnected, the package pin allocations are quite different, and Series 54/74 devices can not be used to replace S.U.H.L. (and vice versa). The Flat-Pack Series 54/74 devices have quite different pin allocations from the Dual-in-line package versions, and care must be taken to use the correct pin allocation chart when boards are being laid out.

Series 54 is the full temperature range, military, version of the family, and Series 74 is the restricted temperature, commercial, version. There is

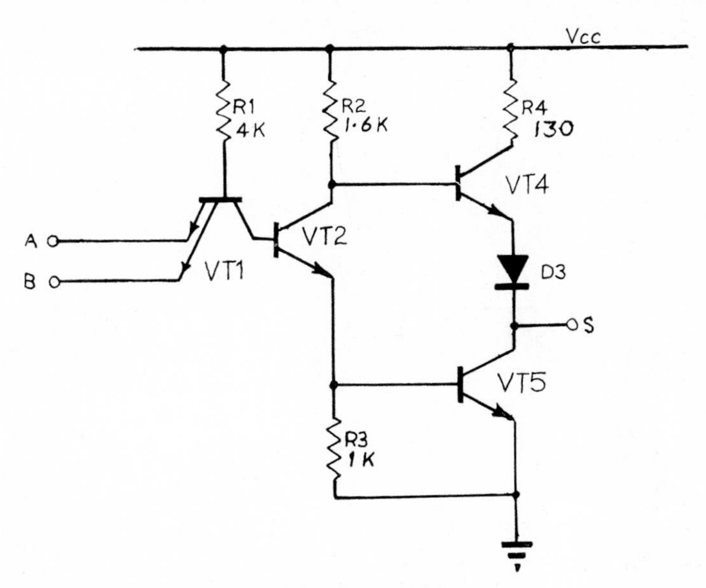

Fig. 3.8 Basic series 74 circuit.

also a '64' range with an intermediate working temperature range. Texas Instruments also market Series 10 and System 11 T.T.L.s which are similar to Series 54/74. (See manufacturers' specifications for full details.)

3.2.3 THE SERIES 9000 GATE

The third major T.T.L. family is the Series 9000 Gate, which was introduced by S.G.S.-Fairchild as a member of their C.C.S.L. (Compatible Current-Sinking Logic) family. The 9000 gate is slightly faster than S.U.H.L. 1, and has comparable power dissipation. The fan-out is 10, and the input leakage current is specified at a value between those for Series 54/74 and S.U.H.L. 1. The circuit is shown in Fig. 3.9. It can be seen that D3 has been replaced by a transistor VT3, which gives a better

'1' level drive capability and improves the switching speed on 'turn-off'.

Current levels in R1 are the same as for S.U.H.L. 1 and Series 54/74, and the currents in R2 are similar to those for Series 54/74. In the ON state, VT3 and R6 pass about 0·025 mA and the nominal base drive to VT5 is 2·86 mA. The OFF state current consumption is nominally 1·08 mA and that for the ON state is 3·19 mA.

Fairchild, S.G.S., Marconi-Elliott Microelectronics, and I.T.T. Semiconductors are the main suppliers of Series 9000. Series 9000 devices generally follow the same range of gates and package pin allocations as Series 54/74, so most of the gates can be used as a faster replacement for Series 54/74 devices when necessary.

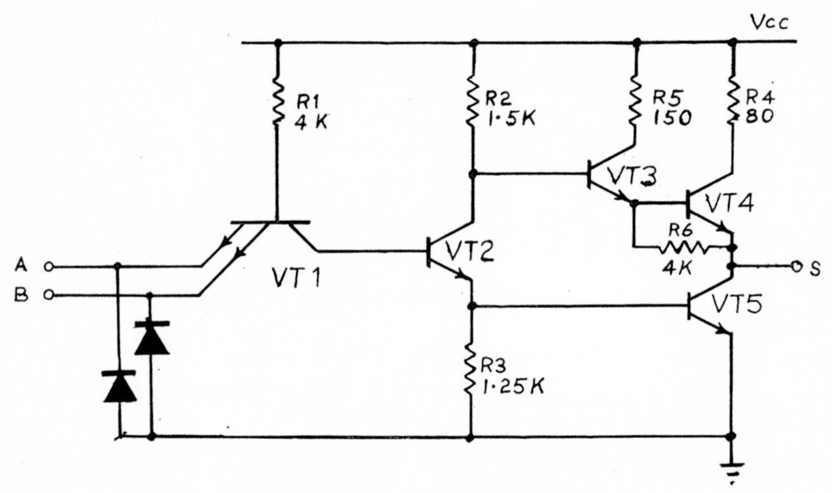

Fig. 3.9 Basic Series 9000 circuit.

3.2.4 THE S.U.H.L. 2 GATE

S.U.H.L. 2 is the high speed member of Sylvania's T.T.L. family. It is faster than Series 9000, has a higher power dissipation, and, like S.U.H.L. 1, is available in high and low fan-out versions. The circuit is shown in Fig. 3.10. As in the 9000 circuit, D3 is replaced by a transistor, but R6 is returned to ground instead of to the collector of VT5. All resistor values are lower than for S.U.H.L. 1 or 9000 gates, and the input leakage current is specified as a maximum of twice that of S.U.H.L. 1.

The currents in R1 are increased to 1·63 mA in the OFF state and 1·12 mA in the ON state. In the OFF state, R6 passes 1·2 mA, and the nominal supply current is 2·1 mA. In the ON state, R2 passes 5 mA, of which 3·6 mA is available as base drive to VT5, and the supply current is 6·14 mA.

The range of gates is similar to the S.U.H.L. 1 range, with full package pinning compatibility. It is believed that in some cases the same diffusion masks are used for S.U.H.L. 1 and S.U.H.L. 2. Different doping levels give the lower value resistors required for S.U.H.L. 2, and different metallization patterns connect VT3 to R5 and R6 for S.U.H.L. 2, or as a diode for S.U.H.L. 1.

RAY 2 and M.T.T.L. 2 are the same as S.U.H.L. 2.

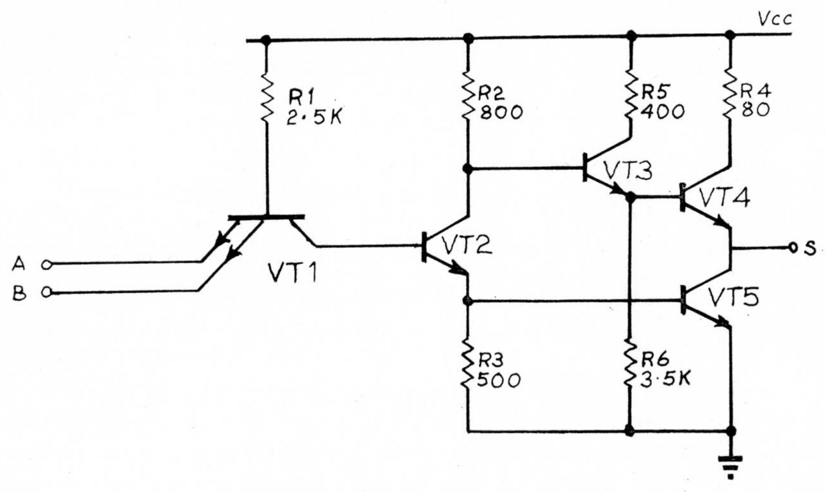

Fig. 3.10 Basic S.U.H.L. 2 circuit.

3.2.5 THE H.L.T.T.L 2 GATE

Transitron Electronic also offer a high speed S.U.H.L. compatible family, but their circuit differs from the S.U.H.L. 2 circuit in that VT3 is connected as a Darlington pair with VT4 instead of having its own collector resistor (see Fig. 3.11). As with H.L.T.T.L. 1 and S.U.H.L. 1, there are minor differences between the specifications for H.L.T.T.L. 2 and S.U.H.L. 2, but the devices can generally be regarded as being fully compatible, possibly with a slight loss in fan-out. (See Section 9.3.)

3.2.6 THE 54H/74H GATE

Series 54H/74H is the equivalent to a 'Series 2' family made by Texas Instruments. The circuit is as for the H.L.T.T.L. 2 gate (Fig. 3.11), but the nominal values of the resistors differ slightly from those of S.U.H.L. 2. R1 is increased to 2·8 KΩ to give a slightly lower input current; R2 is decreased to 760 Ω, and R3 to 470 Ω. The biggest difference is R4, which is decreased to 58 Ω. R6 is increased to 4 K.

The range of gates available follows the standard Series 54/74 range.

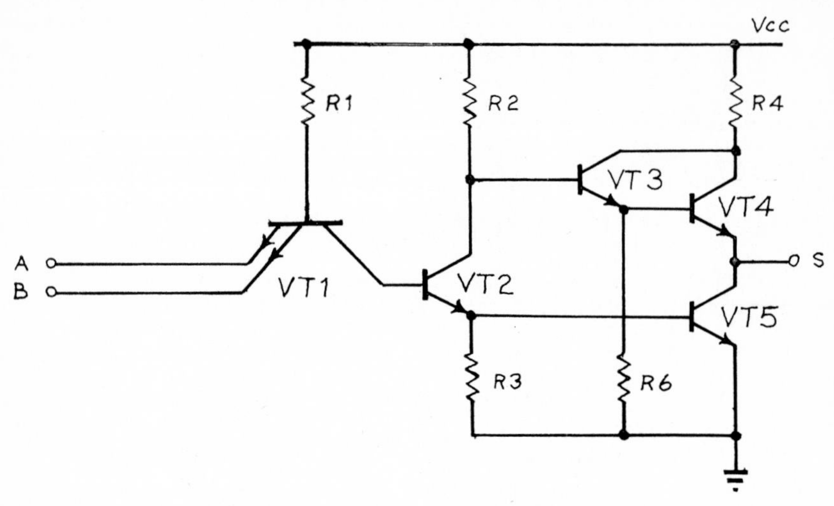

Fig. 3.11 Basic H.L.T.T.L. 2 circuit.

3.2.7 THE RAY 3 GATE

RAY 3 is an even higher speed logic family than RAY 2 or S.U.H.L. 2, but it has the same power dissipation. Improvements in the manufacturing processes have enabled clearances on the chip to be reduced and components to be made smaller, so for the same current levels faster devices can be made. The basic RAY 3 circuit is the same as the M.T.T.L. 3 circuit shown in Fig. 3.12. RAY 3 is pin compatible with S.U.H.L. 1 and S.U.H.L. 2. So far as is known at the time of writing, it is not second sourced.

3.2.8 THE M.T.T.L. 3 GATE

M.T.T.L. 3 is Motorola's 'third generation' T.T.L. The specified speed is the same as that of S.U.H.L. 2, but unlike M.T.T.L. 1 and 2 and RAY 3, M.T.T.L. 3 follows the Series 54/74 package pinning.

The circuit is shown in Fig. 3.12. The main difference between this circuit and the '2' families of other manufacturers is the incorporation of R7 and VT6, which affect the low frequency transfer characteristic (see Section 4.3).

3.2.9 THE SERIES 54L/74L GATE

Series 54L/74L is low power T.T.L., sometimes called 'Milliwatt T.T.L.'. It was introduced by Texas Instruments, and at a casual glance appears to be the same as Series 54/74 (see Fig. 3.13) until it is noted that

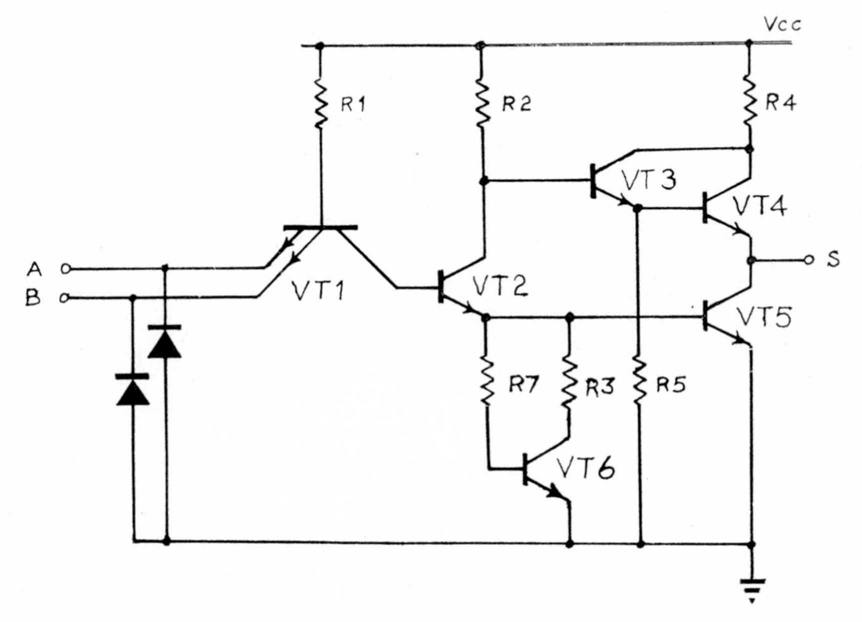

Fig. 3.12 Basic M.T.T.L. 3 circuit.

the resistor values are around ten times those of Series 54/74. This results in a much slower gate (60 ns), but with greatly reduced dissipation, because the ON state current is less than 0·5 mA. This family is really outside the scope of this book, but it is included here for completeness.

3.2.10 THE 54S/74S GATE

The newest member of the T.T.L. range is the Schottky clamped version introduced by Texas Instruments. The basic circuit follows the RAY 3 or M.T.T.L. 3 circuit as shown in Fig. 3.12, except that all the active devices except for VT4 are made with Schottky-barrier diodes across the base–collector junctions.

The diodes are formed by allowing the base metallization to contact the n-type collector region so that it forms a metal to silicon diode to the collector. The Schottky-barrier diode so formed has a lower forward voltage drop than the silicon p–n junction, and it is free from minority carriers. The Schottky-barrier diode prevents the transistor from saturating—current into the base, which would saturate an unclamped transistor and would leave a stored base charge after the current flow into the base ceases, is diverted through the diode. Because the transistors and their Schottky diodes can hold virtually no stored charge, the switching action

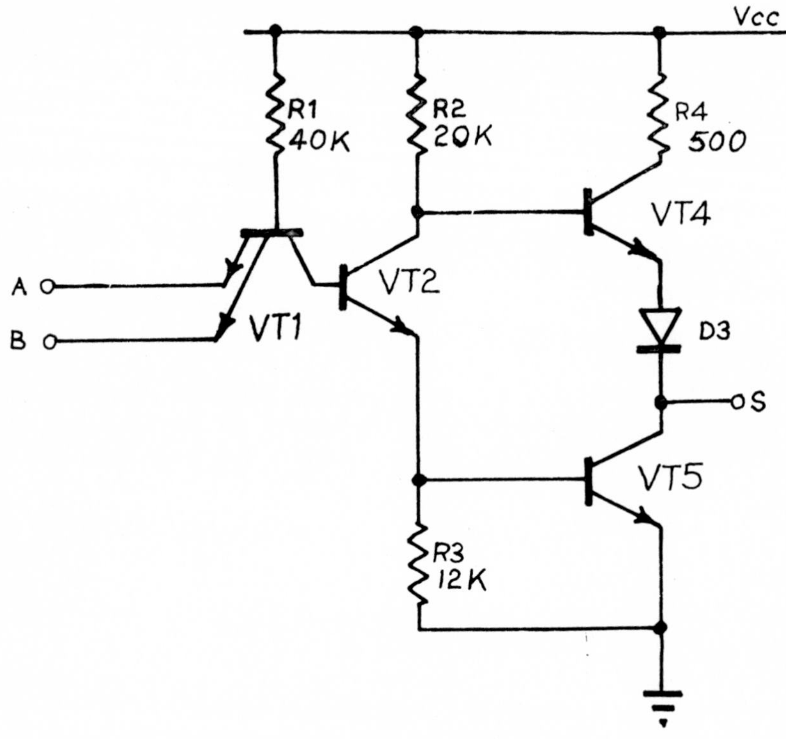

Fig. 3.13 Basic Series 74L circuit.

of the gate is made faster without the customary penalty of increased dissipation.

The input current is specified as 2 mA maximum in the '0' state, with '1' state input leakage current specified as 100 micro-amperes maximum. The Schottky diode across VT5 raises the limit value of V_{OL} to 0·5 V maximum. The dissipation is specified at very nearly the same as that of the 54H/74H devices, but the propagation delay is specified as having a maximum value better than half that of the 'H' gates. The gain in speed is almost entirely caused by the absence of stored charges—the switching speeds and hence the edge speeds of the output are comparable to those of 54H/74H, and so the faster 'S' devices impose no tighter interconnection constraints than do the 'H' devices.

The virtual absence of stored base charge in VT5 means that the 'switching spike' is considerably smaller than that for any other Series 2 or Series 3 devices (see Chapter 6). The Schottky-barrier diodes on the input pads clamp the inputs at a lower negative voltage than the conventional diodes on other types of gates, and they also permit faster working.

4

DC and Low-frequency Characteristics

4.1 Input Characteristic

The input characteristics of all the families of T.T.L. gates are very similar, because there is little or no difference in the input circuits used. Fig. 4.1 shows a typical input characteristic. With 0 volts applied to an input, current flows out of the gate through R1 and VT1, so the value of the current is set by the value of R1. As the applied voltage is raised the current falls linearly until the switching threshold is reached, when, as the current through R1 splits between the emitter and collector of VT1, the input current falls sharply to zero. The re-entrant portion of the characteristic at the start of switching shown on the larger curve in Fig. 4.1 has been observed on a number of gates from several different manufacturers. In the drawing it is exaggerated slightly. It is believed to be caused by stored base charge effects in VT1, and its only practical effect is to sharpen slightly the actual switching when a fast-edged pulse is applied to the input.

Above the switching threshold, leakage current flows into the input. This current is not the normal leakage current of an OFF transistor, but is considerably higher. The high leakage current is caused by inverse transistor action as the current through R1 flows to VT2 through the collector of VT1. In Fig. 4.2, VT1 has been re-drawn with only one emitter and with the emitter arrowhead on the collector to make this action clear. This input leakage current is usually called the Inverse Beta Current (I_BIN). If other inputs to the same AND gate are earthed, transistor action can also occur between the high and low emitters ($I_{IN(LK)}$ or I_{IH}).

The input transistors are designed to minimize these leakage currents. Base diffusions are usually gold doped to reduce the lifetime of the minority carriers, and the geometry of the input transistors is usually such as to partially isolate the multiple emitter areas. Fig. 4.3 shows a heart-shaped transistor used by one major manufacturer. Other manufacturers use a cruciform geometry to achieve the same object.

All manufacturers appear to be very successful in reducing this undesirable input leakage current as the average values measured by the author and his associates have been well below the specified limits. The

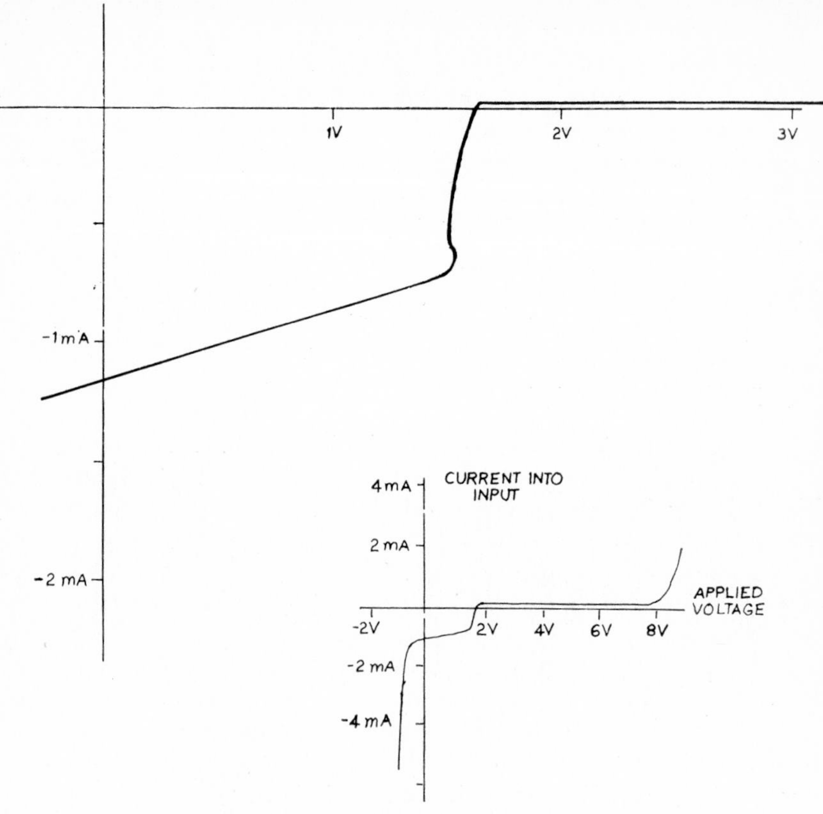

Fig. 4.1 Typical T.T.L. input characteristic.

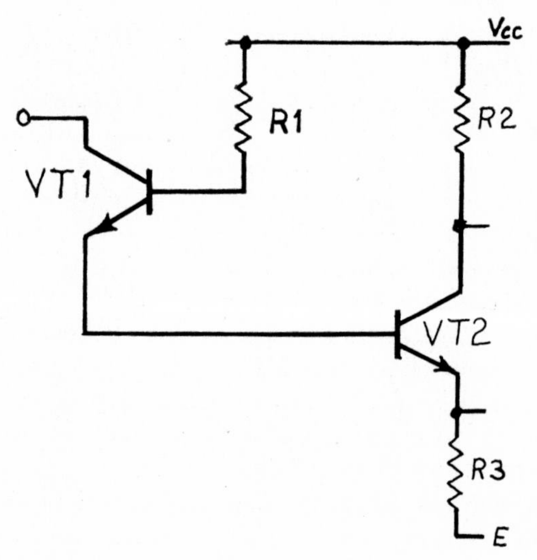

Fig. 4.2 Input transistor inverted to show inverse beta action.

'1' level leakage current is fairly constant from the switching threshold up to about 4 V, but above 4 V it rises until breakdown occurs. All manufacturers rate the gate inputs at 5·5 V absolute maximum, and warn that inputs can be damaged if this figure is exceeded. On the gates tested to destruction by the author, input breakdown did not occur until about 8 V but cases are known in which inputs have been damaged because they have been returned to a 5 V supply rail which has 'spiked' to above 5·5 V. (See Section 9.2.5.)

If an input is taken below earth, the normal current through R1 flows

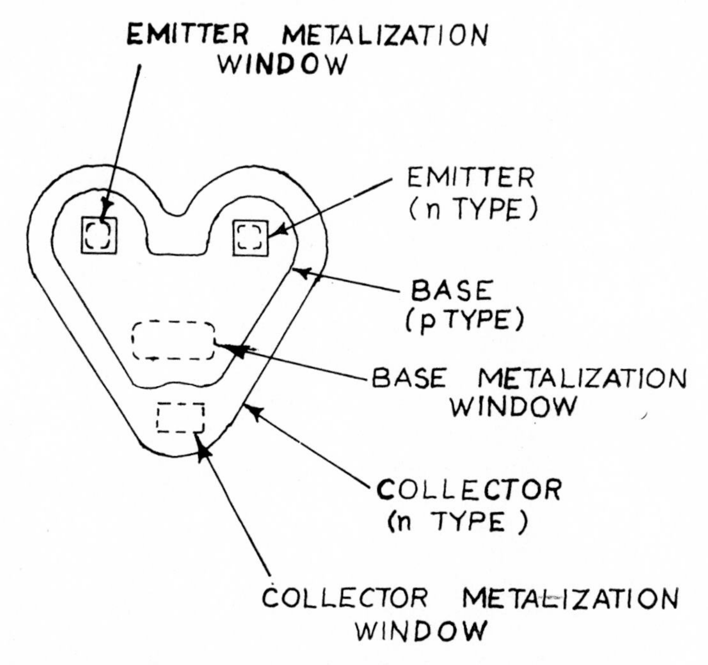

Fig. 4.3 Multi-emitter transistor pattern.

out until the voltage on the input has fallen to a level at which the collector–substrate diode of VT1 becomes forward biased. In Fig. 4.4, VT1 has been drawn below the earth rail, with its collector–substrate diode shown, and it can be seen that normal transistor action will cause a current of $I_{R1}h_{FE}VT1$ to flow out of the input. This action limits the negative excursion of the input, and is of great value in reducing the effects of ringing when gates are driven from long tracks or transmission lines. Some manufacturers design VT1 to exhibit this effect with a fairly sharp cut-off to the negative excursion at about —1 V, but other manufacturers prefer to leave the collector–substrate diode impedance fairly high and diffuse separate diodes beneath the input bonding pads to limit the possible negative excursion of the inputs (see Figs. 3.9 and 3.12).

The application of negative voltages, from low impedance sources, to inputs should be avoided as very high, possibly destructive, currents can flow. Two manufacturers have stated that a current of 10 mA out of an input terminal is the maximum that should be allowed.

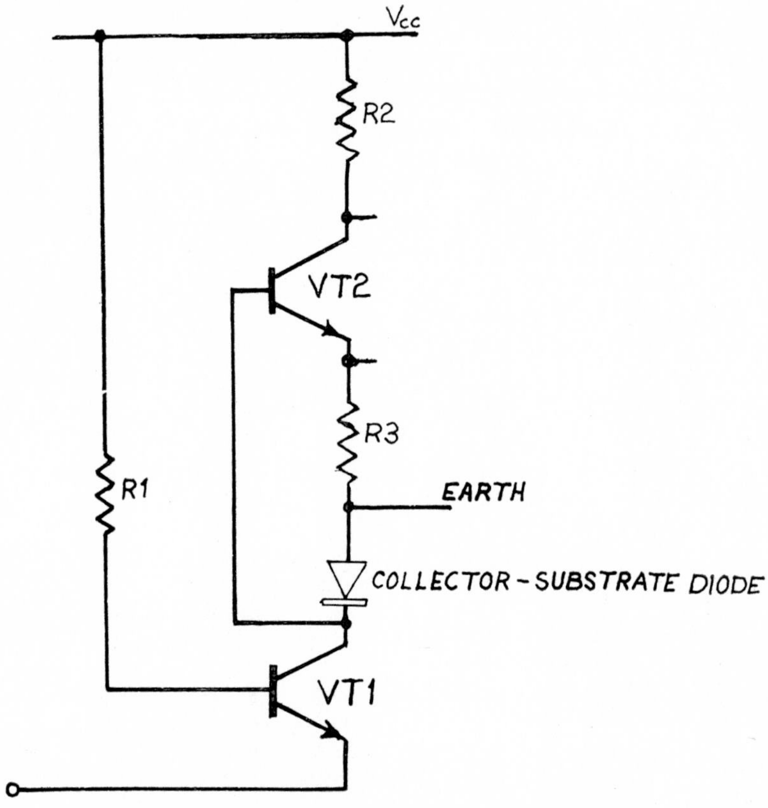

Fig. 4.4 Input transistor showing substrate diode.

4.2 Output Characteristic

The normal output voltage swing of a T.T.L. gate is typically between 0·2 V in the ON state and 3·5 V in the OFF state. As different parts of the circuit are active in the two states, there are two different output characteristics as shown in Fig. 4.5. The differences between the output circuits used for the different families of T.T.L. result in minor differences in the high level output characteristic. The effects of these differences are explained in Section 14.3.

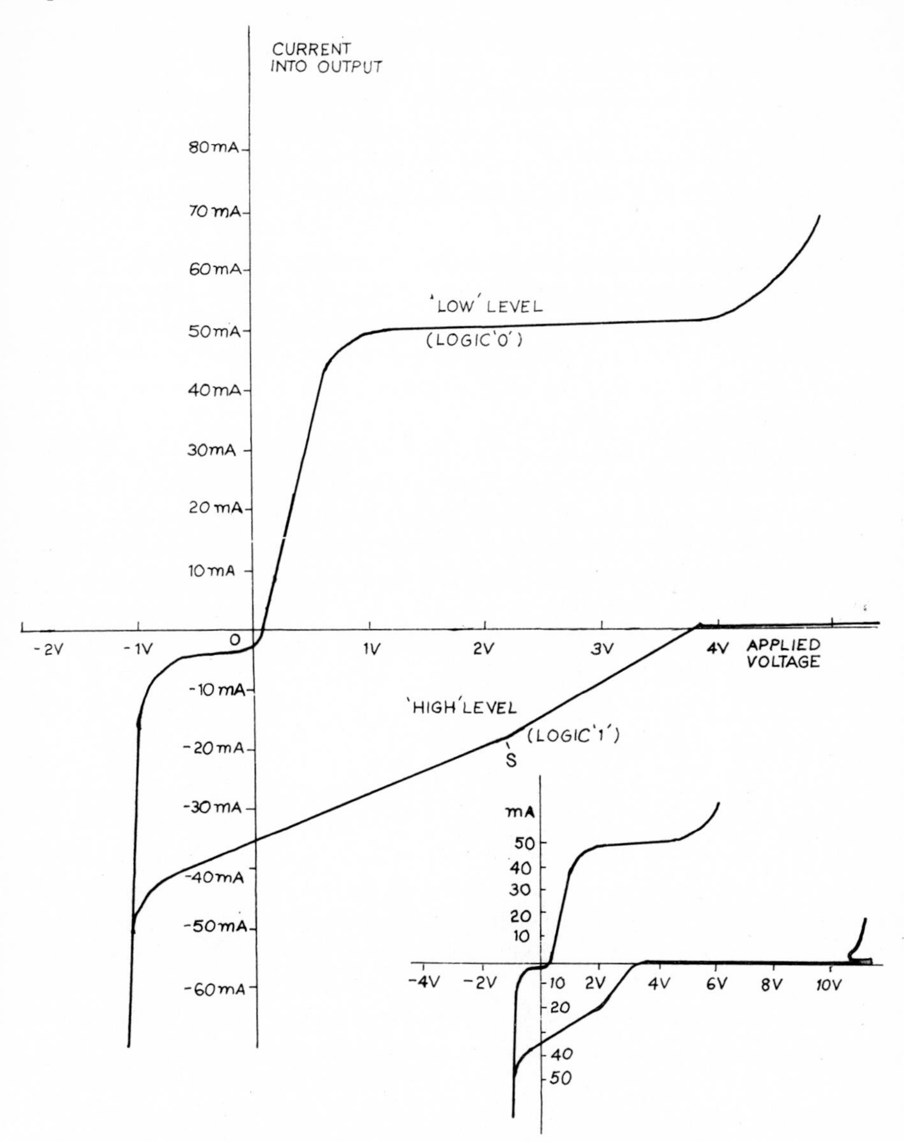

Fig. 4.5 Typical T.T.L. output characteristic.

4.2.1 LOW LEVEL CHARACTERISTIC

The area of the low level characteristic of most interest is that between 1 mA and 20 mA current into the output. This portion of the characteristic shows the normal output impedance of a saturated transistor, with an impedance of around 10 Ω. On all gates measured VT5 remained in saturation up to around 40 mA, although the voltage had sometimes risen above the specified value of V_{OL}. If a high voltage is applied to an ON output the current rises sharply above about 4 V. As testing at this level is liable to damage the gates only a few samples were tested.

At the specified full load output currents all gates tested were well within the manufacturers specifications, and 0·2 V was the typical V_{OL} found.

When the voltage applied to an output is taken below the normal value of V_{OL}, current into the output ceases at $V_{CE(sat)}$VT5, and as the output is taken further negative, current flows out of the output. This current is the base drive to VT5 from VT2, and it flows through the forward-biased base–collector diode of VT5 until at about −0·5 V all the base current is flowing out via the collector.

As the output potential is taken further negative, the collector–substrate diode of VT5 begins to conduct. Current out of the output then increases rapidly and the negative excursion of the node is limited to about −1 V. No current can flow through VT4, because VT2 is saturated, and the base to collector drop across VT5 does not rise to a sufficient level to turn on D3 (or VT3) and VT4. The slope of the low level output characteristic between V_{OL} and −0·5 V is the most important factor in the determination of line ringing (see Section 14.3).

4.2.2 HIGH LEVEL CHARACTERISTIC

As was explained in Section 3.1.3, the normal OFF output voltage is two semiconductor drops below the supply rail voltage, i.e. with a 5·0 V rail, V_{OH} is about 3·5 V. Since T.T.L. inputs sink current at voltages above V_{TH}, a '1' level output must be capable of sourcing current. This current is supplied by VT4 acting as an emitter follower. At low currents (less than 0·5 mA) the voltage drop across R4 will not be equal to the difference between V_{BE}VT4 and V_{CE}VT4, so VT4 will not be saturated and the output impedance will be R2/h_{FE}VT4, which will be around 80 Ω. As the output current rises, the voltage drop across R4 rises until at point S on the high level output characteristic VT4 saturates and the output impedance changes to approximately that of R4. A high level output can be shorted to earth, when I_{OS} of around 40 mA will flow. If the output is taken below earth, the current flow increases until, as in the case of the '0' state output, the collector–substrate diode of VT5 starts to conduct and limits the negative voltage excursion.

The point on the characteristic at which the output impedance rises is influenced by whether D3 is on the base or the emitter of VT4. With the diode on the base, I_{OH}R4 must rise by

$$(V_{BE}\mathrm{VT4} - V_{CE}\mathrm{VT4}) + V_{D3} + I_{OUT}\mathrm{R2}/h_{FE}\mathrm{VT4}$$

before VT4 can saturate, whereas with the diode on the emitter, V_{D3} can be ignored. Output circuits with two transistors show differences in the high level output characteristics depending on whether or not VT4 can saturate, and whether it saturates before or after VT3.

The presence of R6 across the base-emitter of VT4 on some types of gates affects the low current output characteristic. At very low currents (less than 0·05 mA) this resistor will cut off VT4, and the normal output level will be only one semiconductor drop below V_{CC}. As the output current rises, the voltage drop across this resistor rises until at less than 0·2 mA VT4 is acting as an emitter follower and the output characteristic follows that for gates without this resistor. (See Fig. 4.6.)

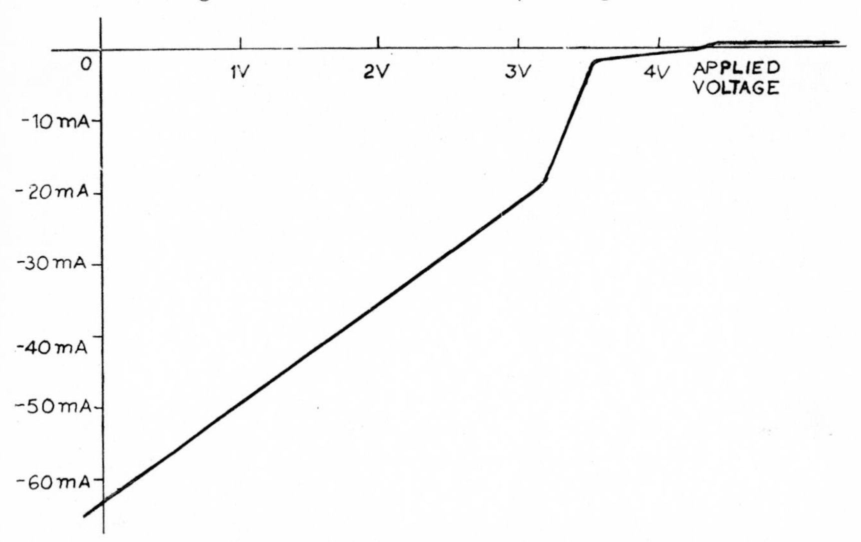

Fig. 4.6 High level output characteristic of gate with R6 across VT4.

If voltages in excess of V_{OH} are applied to a high level output, VT4 cuts off and only negligible leakage current flows into the output until breakdown (presumably of VT5 collector) occurs at around 10 V and the current increases sharply.

4.3 Transfer Characteristic

The typical low frequency variation of output voltage with input voltage of a T.T.L. gate is shown in Fig. 4.7. The input voltage may be considered as being applied to all inputs to the gate. If it is applied to one input only, the other inputs must be returned to a logical '1' level.

While the input is at 0 volts, the output will be at about 3·5 V. The base of VT1 is at V_{BE}VT1 above earth, and VT2 and VT5 are cut-off.

The base of VT5 will be at earth, and that of VT2 will be at or near earth. As the input voltage is raised, the base of VT1 rises, keeping V_{BE} above the applied voltage. The collector of VT1 and the base of VT2 also rise, following V_{CB} below the base of VT1, i.e. the base of VT2 rises at about the same level as the applied voltage.

When the applied voltage rises to about 0·65 V, at point A on Fig. 4.7, the base of VT2 will be at roughly the same voltage and VT2 will begin

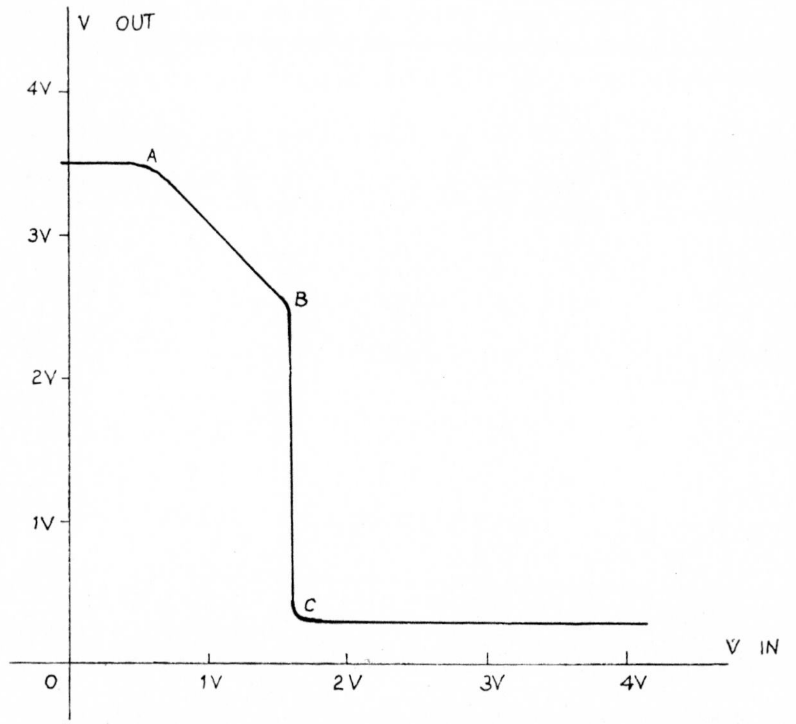

Fig. 4.7 Typical low frequency transfer characteristic of T.T.L. gate.

to conduct. The emitter of VT2 will rise, following at about V_{BE}VT2 below the applied voltage. The voltage on the emitter of VT2 causes current to flow through R3, and this current must come mainly through R2 (for this discussion the base current into VT2 can be ignored). The values of R2 and R3 are similar, and in some families they are the same, so that as the emitter of VT2 rises the collector will fall by roughly the same amount. Under normal load conditions, the output voltage will be $V_{D3} + V_{BE}$VT4 below the voltage on the collector of VT2, so that over the portion of the transfer characteristic above about 0·7 V V_{IN} the output voltage will fall at roughly the same rate as the input voltage rises, until at B on Fig. 4.7 the voltage on the emitter of VT2 will have risen high enough to turn on VT5. This occurs when the input voltage has risen to about 1·4–1·5 V. The current flow through VT2 increases rapidly, as it is no longer limited by R3, until VT2 saturates at about 4 mA current. R3 passes only about 0·7 mA, and the remainder of the current through VT2 provides base drive to VT5, which turns hard on so that the output voltage changes to V_{OL} at point C on Fig. 4.7.

The sloping portion of the characteristic between A and B implies that

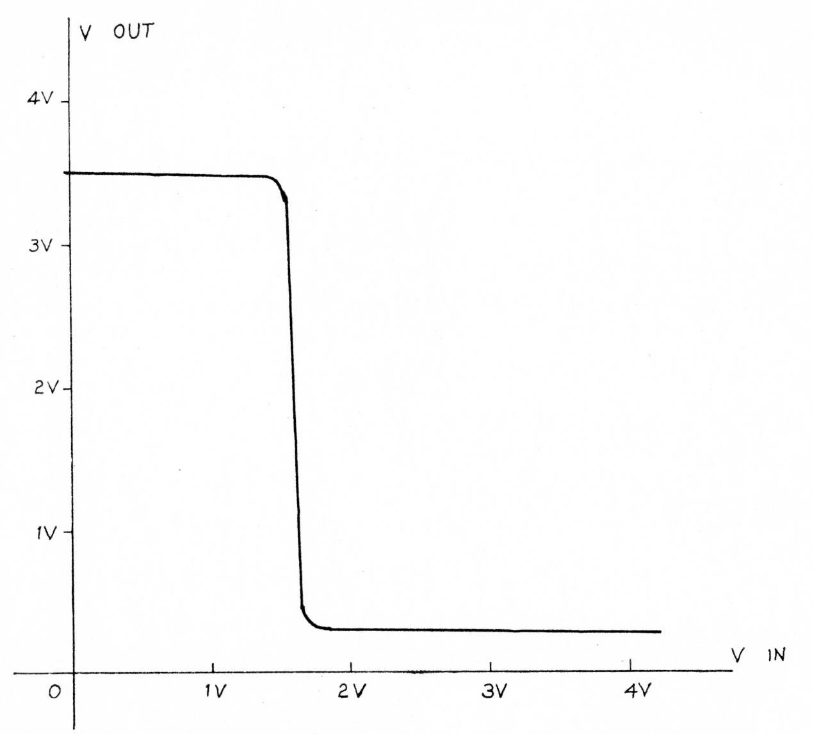

Fig. 4.8 Transfer characteristic of gate with R7 and VT6 added.

any noise between 0·7 and 1·4 V present on the '0' level input signal should appear inverted and at roughly the same amplitude on the '1' level output signal. Chapter 9, Section 4.5 explains the practical significance of this sloping portion of the transfer characteristic.

Some of the later T.T.L. circuits replace the single resistor R3 with a resistor–transistor network (see Fig. 3.12). When this circuit is used, no current flows through VT2 until the voltage on the emitter of VT2 has risen to above V_{BE}VT6, and then R3 can pass current. Since V_{BE}VT6 will be roughly equal to V_{BE}VT5, the '1' level output will remain undisturbed until the input voltage has risen to the point where VT5 turns on.

This means the slope between A and B in the normal transfer characteristic is eliminated, and the gate with R7 and VT6 added has a transfer characteristic as shown in Fig. 4.8. This type of gate rejects low frequency noise at any level below the actual switching threshold.

5

Switching of the T.T.L. Gate

5.1 Simplified Switching Action—Turn-on

When a T.T.L. gate is switched by another T.T.L. gate to which it is connected by only a short wire or track, the change of voltage on the input pin is very fast (about 0·3 V/ns). (See Fig. 5.1.) As the voltage on the input (emitter of VT1) rises, the base of VT1 rises with it, at V_{BE} higher voltage, and the current sourced by the input falls. The collector of VT1, which is connected to the base of VT2, follows the base of VT1, and begins to pass current through the base-emitter of VT2 at about 0·6 V. The current through the emitter diode of VT2 and R3 causes the voltage on the emitter of VT2 to rise until the base-emitter diode of VT5 starts to conduct. The base–emitter diode of VT5 limits the rise of VT2 emitter to about 0·7 V, possibly after a small overshoot, and V_{BE}VT2 limits the rise of VT1 collector to about 1·4 V, which in turn limits the rise of the base of VT1. The input potential continues to rise, cutting off the base–emitter diode of VT1, and leakage current (or rather reverse transistor current) is sunk by the input.

About 4 ns after base current begins to flow in VT2, collector current will start to flow through R2. The rate of fall in potential of VT2 collector is governed by the rate of rise of the input voltage, the f_T of VT2, the value of R2, and the stray capacity present. Diode D3 and the base-emitter of VT4 have been passing current (and current has also been flowing through VT4 collector–emitter), but as the collector of VT2 falls in potential, VT4 is cut-off and the emitter of VT4 (the output of the gate) falls very slowly on a long time constant governed by the current which is sunk by the input of the succeeding gate (I_{IH}) and the capacity on the output node.

Collector current begins to flow in VT5 about 4–5 ns after its base reaches about 0·7 V. Initially this current has only to discharge the capacity on the output node, and the rate of fall of voltage of the output node is governed by the f_T of VT5 and the capacity present. Below about 1·5 V VT5 has to sink the load current from the input of the succeeding gate, and the rate of change of voltage decreases slightly.

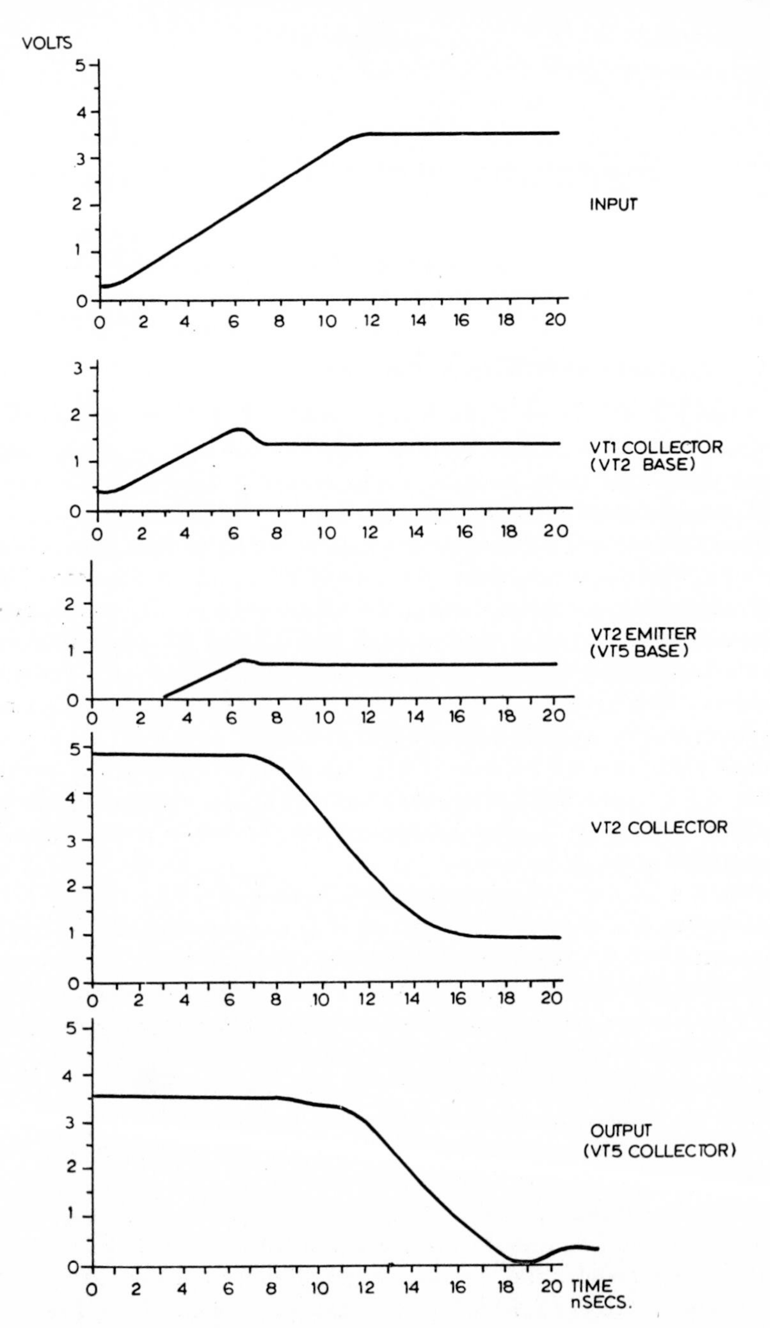

Fig. 5.1 Typical T.T.L. turn-on action.

On a simple gate VT2 turns on completely before VT5 can complete its turn-on, but on gates with internal OR functions the extra capacity on the collector of VT2 can slow down the fall in potential at this point until at some time during the turn-on, the collector of VT5 reaches a potential more than 1·4 V below that of the collector of VT2. D3 and VT4 will then conduct, and the rate of change of the output voltage will be governed by the turn-on of VT2. While VT4 and VT5 are both conducting, current can flow throuth R4, VT4, and VT5. This current, which is usually not more than a few milliamperes, ceases when the collector of VT2 has fallen sufficiently to turn off D3 and VT4.

5.2 Simplified Switching Action—Turn-off

During turn-off of a T.T.L. gate substantially nothing happens until the input potential has fallen to the threshold voltage. (See Fig. 5.2.) As the potential on the input reaches a level approximately equal to that on the collector of VT1, current begins to flow out of the input terminal, stopping the current flow through VT1 collector to the base of VT2. As the voltage on the input continues to fall, VT1 is biased as a normal transistor with forward gain, and collector current flows out of the base of VT2, withdrawing the stored charge from VT2 and turning it off quickly. The base of VT2 follows the input potential down with a slight time lag as the stored base charge is withdrawn.

The voltage on the emitter of VT2 falls as VT2 turns off, but it is held after only a slight fall by the base of VT5, since VT5 is still conducting and still has stored base charge. The collector of VT2 starts to rise as VT2 turns off, but after it has risen about a volt, D3 and VT4 turn on, since the collector of VT5 is still held low by the stored base charge in VT5. The collector voltage of VT2 is then held down by the current flow through R2, D3, VT4, and VT5. As VT4 turns on, current flows through R4, VT4, and VT5. The amplitude of this current is limited by the value of R4 to a maximum of about 45 mA. In practice, this current is typically about 8 mA. The current flow through VT5 sweeps out the stored base charge until VT5 turns off and the voltage on its collector rises, allowing the collector of VT2 to rise. The base potential of VT5 then falls towards zero on a time constant which is controlled by R3 and the stray capacity.

As the voltage on the collector of VT5 rises, VT4 remains saturated and the current through R4, VT4, and VT5 decays. This 'spike' of current through the output stage is characteristic of T.T.L. devices with active pull-up circuits. The amplitude of the current is limited well below the d.c. output short circuit current by the time required to establish the current flow, and its duration is limited by the turn-off time of VT5.

The output potential rises until VT4 cuts off, often with an overshoot

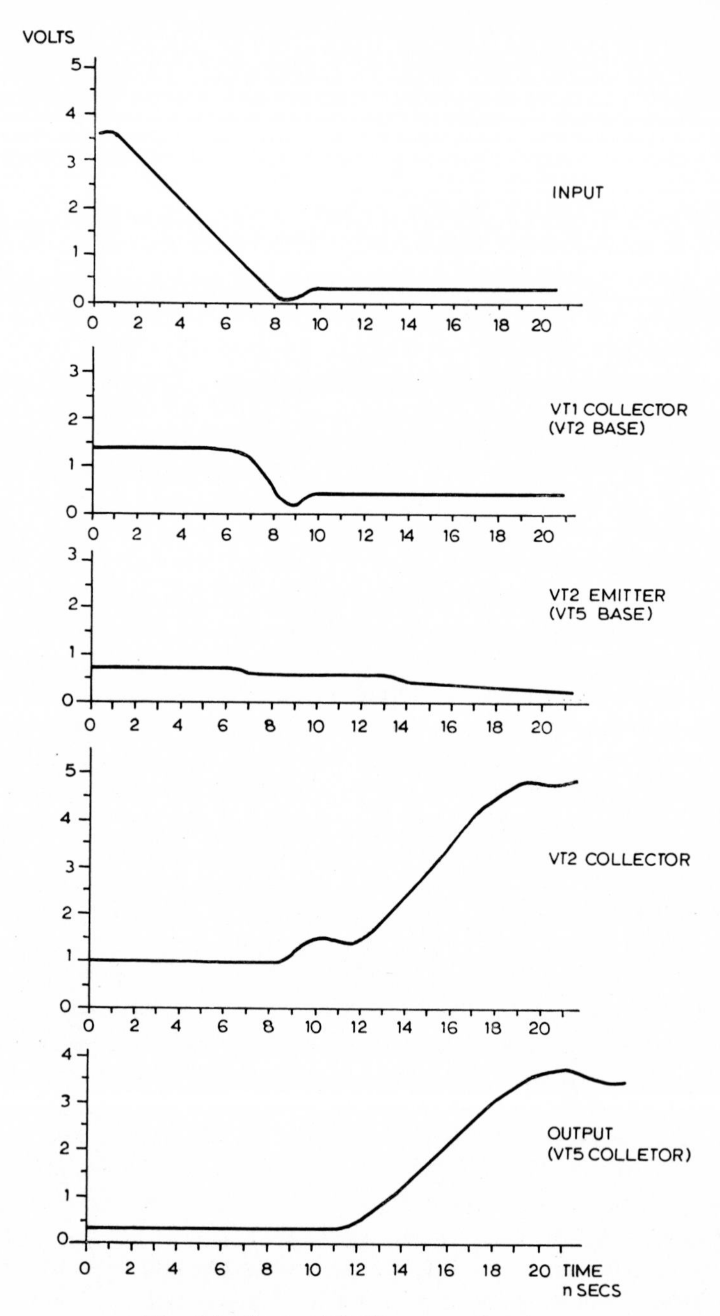

Fig. 5.2 Typical T.T.L. turn-off action.

caused by line reflections. (See Section 14.3.2.1.) This overshoot leaves the output with a high impedance source and a high impedance load at the input to the next gate, so recovery to a normal '1' level with VT4 on the verge of conduction may be prolonged.

5.3 General Effects of Switching

5.3.1 CHANGE OF PROPAGATION DELAY WITH FREQUENCY

It can be seen from the description of the switching sequence of events that the circuit might not be completely restored to a normal d.c. condition immediately after the output has switched. This could result in the propagation delay of a device varying slightly as its frequency of switching is varied. Tests have shown that between 2 MHz and 20 MHz the greatest change in propagation delay should not exceed half a nanosecond. The direction of the change, i.e. speed-up or slow-down with increasing frequency, was found to be unpredictable.

5.3.2 CHANGE OF DEVICE DISSIPATION WITH FREQUENCY

A serious effect of rapid switching is an increase in device dissipation. As is described in Section 5.2 a surge of current is drawn from the supply rail through R4, VT4, and VT5 during turn-off. The magnitude and duration of this surge are substantially independent of the frequency of switching. At low frequencies this surge has little effect on the average power consumption of the gate, but as the frequency of switching is raised these surges account for an increasing percentage of the power consumption, until at 20 MHz a gate may be taking as much as four times its average low frequency supply current (see Fig. 5.3). The ratio of average power consumption at high frequency to average d.c. power consumption varies widely from gate to gate as it depends on the normal d.c. power consumption and on the magnitude and duration of the current spike, which in turn depend on the loading on the gate. Fig. 5.3 shows five measured curves and a predicted curve (dotted) for a worst-case spike on a gate with the maximum allowable d.c. dissipation.

5.3.3 EFFECTS CAUSED BY SWITCHING OF TWO GATES ON THE SAME CHIP

When several gates are formed on the same chip the simultaneous independent switching of two or more gates can have a very small effect on the speeds of the gates. Like the effects of the frequency of switching on the delay, this effect varies from package to package.

The switching of one gate in a multi-gate package can affect what should be a steady '0' level output from another gate in the same package. The

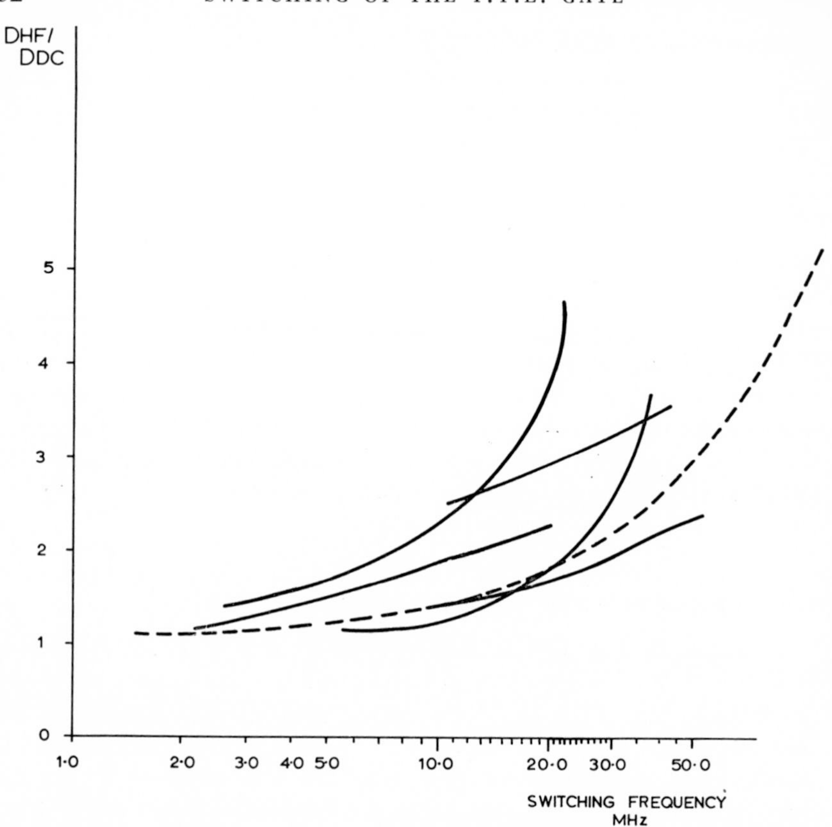

Fig. 5.3 Ratio of device dissipation at high frequency to average d.c. dissipation.

current which is sunk by all ON outputs is returned to the board or system earth through a common earth return path on the chip, and through the bonding wire to the lead frame. As gates are turned on and off the current flowing in this common path will change, and although the resistance of the common path is low, it is not completely negligible and so changes in output voltage levels must be expected. These changes are likely to be significant only on heavily loaded line drivers. So far as is known to the author, all manufacturers test multi-output devices with all outputs fully loaded, so any changes in '0' level voltages caused by common earth path effects should all occur at levels below the specified low output voltage level.

6

The T.T.L. Switching Spike

6.1 The Current Spike

6.1.1 INTRODUCTION

The current spike which occurs in the supply rail when a T.T.L. device switches has been regarded by many engineers as the biggest disadvantage of T.T.L., and has probably led to more misconceptions than any other feature of any logic family. Some engineers have stated (without any justification) that the use of boards with internal earth planes is essential if T.T.L. is to be used; or that unless every package has its own decoupling capacitor the rail spike generated when one gate switches will cause other gates on the board to switch spuriously. The fallacies in such statements are revealed when the rail spike and its effects are given proper consideration.

6.1.2 GENERATION OF THE CURRENT SPIKE DURING TURN-OFF

Section 5.2 explains how, during turn-off, VT4 is turned on while VT5 is still conducting, which results in a 'spike' of current which flows until VT5 turns off.

The supply current to the device, I_{CC}, rises rapidly from the steady state ON level during the turn-off period. As VT2 is turned off and its collector starts to rise, D3 and VT4 turn on. The emitter of VT4 is held at V_{OL} by the collector of VT5, so there is no change in the charge on the capacitance of the output node and the rate of rise of current is limited only by the characteristics of VT4. (Zero source impedance on the supply to the chip is assumed.)

If VT5 remained saturated this current would rise to the output short circuit current level, limited only by R4. However, before D3 and VT4 can begin to conduct VT2 must have turned off, leaving only stored charge in the base circuit of VT5 to maintain conduction. The rapidly rising current through R4, VT4, and VT5 sweeps out this stored base charge and VT5 desaturates, which leaves the current flow limited by the 'active region' collector-emitter resistance of VT5. The output potential rises, with the rate of rise dependent on the node capacitance which has to be charged by the current through VT4. As this capacitance is charged the current flow through VT4 and R4 decreases until I_{CC} has dropped to its steady state OFF value. Some typical current spikes are shown in Fig. 6.1, all of which are drawn with I_{CC} on taken as the zero current level.

Some idea of the likely magnitude of the current spike can be gained

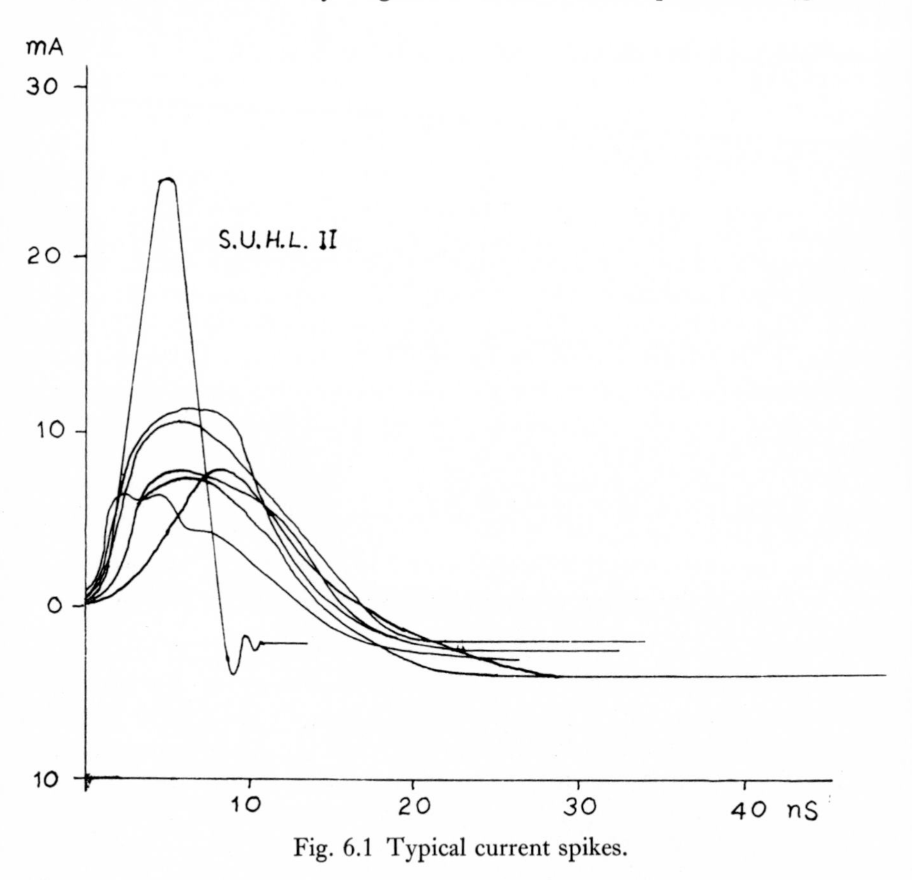

Fig. 6.1 Typical current spikes.

from the characteristics of the circuit. A gate which has a fast turn-off action will probably have a higher amplitude current spike than a gate which is slower, and a gate in which R3 is of low value will tend to have a current spike of short duration, since the stored charge on the base of VT5 can leak away faster.

The author has not measured any current spikes on Series 3 devices (M.T.T.L. 3 or RAY 3), but these are believed to be comparable with those of Series 2 devices (S.U.H.L. 2, H.L.T.T.L. 2). S.U.H.L. 1 devices have a higher amplitude, faster, current spike than Series 54/74 devices, and Series 9000 devices have higher level current spikes than S.U.H.L. 1, but not as high or as fast as S.U.H.L. 2.

The effects of environmental variations (such as temperature) on the current spike are considered in Chapter 10.

6.1.3 GENERATION OF THE CURRENT SPIKE DURING TURN-ON

Section 5.1 explains how capacity on the collector of VT2 on an 'OR—expandable' gate can delay the fall in voltage on the collector of VT2 such that VT5 turns on and its collector falls to a potential below that of the collector of VT2, which allows D3 and VT4 to conduct until the voltage on VT2 collector falls to its normal ON level. This can cause a current spike similar to the turn-off spike, but usually considerably smaller. Under normal conditions of operation this turn-on spike can be ignored.

6.2 Effects Caused by the Current Spike

6.2.1 SUPPLY RAIL VOLTAGE SPIKE

Because it is impossible to achieve a zero source impedance power supply to a device, the current spike will produce a voltage spike on the V_{CC} supply line, relative to the device earth connection. The magnitude, duration, and edge speeds of this voltage spike depend on the nature of the source impedance, which is described in Chapter 16. The magnitude and effects of the voltage spike are discussed in detail in Section 16.2.

6.2.2 INTERFERENCE ON SIGNAL LINES

The sudden changes in current in the supply rail, together with the change in voltage caused by the impedance of the supply rail, can couple into tracks which run adjacent to the supply rail track. Since the voltage swings are considerably less than the normal swing of the output of a gate which is switching (unless the supply rail source impedance is very high), any coupling of a rail spike into a signal line will be small compared with the coupling to be expected between adjacent signal lines, and it can safely be ignored. (Typically, any spurious signal caused by coupling from a supply rail spike can be expected to be between one tenth and one third of the amplitude of a spurious pulse caused by coupling from a signal line.) Chapter 15 describes coupling between adjacent tracks.

7

Variants of the Basic T.T.L. Circuits

Section 3.1 described in detail the basic T.T.L. gate, and its extension to 'AND-OR-INVERT' types by the inclusion of parallel phase-splitter transistors. With other minor changes to the basic circuit the device characteristics can be altered to produce further families of devices. In this section the main variants and their effects on the working of the device are considered. In all drawings, the numbering of the components follows that of Fig. 3.1, instead of going through the circuit from left to right, and once a variant has introduced a new resistor or transistor, etc. the number allocated to that component will be used only for a similarly placed component in any other variant. Thus throughout this section VT4 will always refer to the upper output transistor, R3 to the resistor in the emitter of the phase-splitter, etc.

7.1 The Non-Inverting Gate or 'AND' Gate

The basic T.T.L. gate provides a logical 'NAND' function. There can be occasions when extra inverters have to be used to achieve a desired 'AND' function. This may require extra packages, and it adds to the total delay time of the circuit, so several manufacturers have introduced non-inverting gates which can perform an 'AND' function directly.

The non-inverting gate (see Fig. 7.1) actually comprises a normal inverting gate with an extra inversion stage added, so the full action is 'AND-(OR)-INVERT-INVERT'. The phase-splitter and output stage from VT2 onwards is usually that of a higher speed gate with the addition of R8 from the output to the base of VT5. This resistor keeps VT4 active in the '1' level state and helps to minimize ringing after switching.

The input transistor VT1 follows normal practice, but instead of feeding the phase-splitter, it feeds into the extra inversion transistors VT7 and VT8. Two transistors are necessary at this point to maintain the correct switching level on the inputs, as in a standard gate.

The low frequency transfer characteristic does not have the 'bend' in its upper portion like the normal inverting gate because the phase-splitter

VT2 no longer starts to conduct gradually as the input potential rises, but turns on or off abruptly as VT8 collector passes through the switching threshold of VT2 at about 1·6 V.

The Darlington-pair connection of VT7 and VT8 ensures that VT8 can not saturate, so the collector of VT8 will not fall below about 1 V. This keeps VT2 active at all times, giving a quicker turn-on action. Fig. 7.2 shows the simplified waveforms through the circuit when it is turned off.

Non-inverting gates are generally slower than Series 1 inverting gates in practice, although the limit speeds quoted in the specifications may be the same or even faster, but one non-inverting gate is faster than two

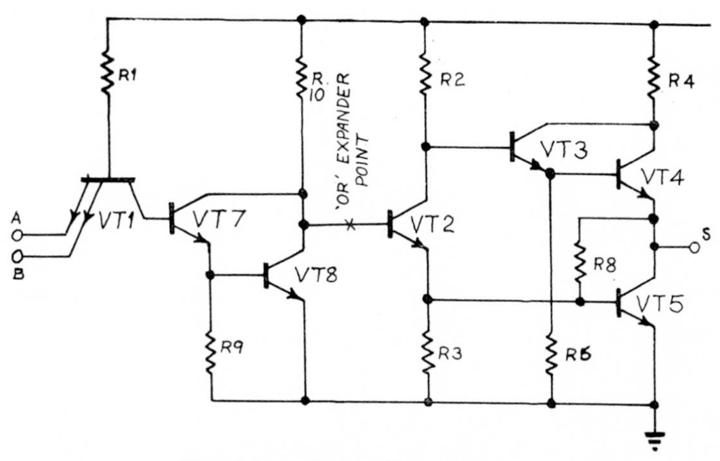

Fig. 7.1 Basic non-inverting gate.

cascaded Series 2 inverting gates. Dissipation is higher than for an inverting gate since either the phase-splitter or the inverting transistors VT7 and VT8 must be on.

Internal 'OR' functions are achieved by the connection of further transistors in parallel with VT7 so that the extra input circuit added is the same as for a normal 'AND-OR-INVERT' gate. (See Fig. 7.3.)

Some non-inverting gates have a 'Transient Control' facility in which the base connection to VT5 is brought out to a terminal, so that a feedback capacitor can be connected across between the collector and the base of VT5, thus varying the output edge speed. With capacitors of about 150 pF, the rising and falling edges can be slowed to about 50 ns. When this transient control facility is provided, VT8 feeds VT2 through a diode with a resistor from the base of VT2 to earth, and the resistor from the output connection to the base of VT5 is replaced by a resistor between the output and earth.

Not all manufacturers who offer 'AND' gates use the Darlington-pai

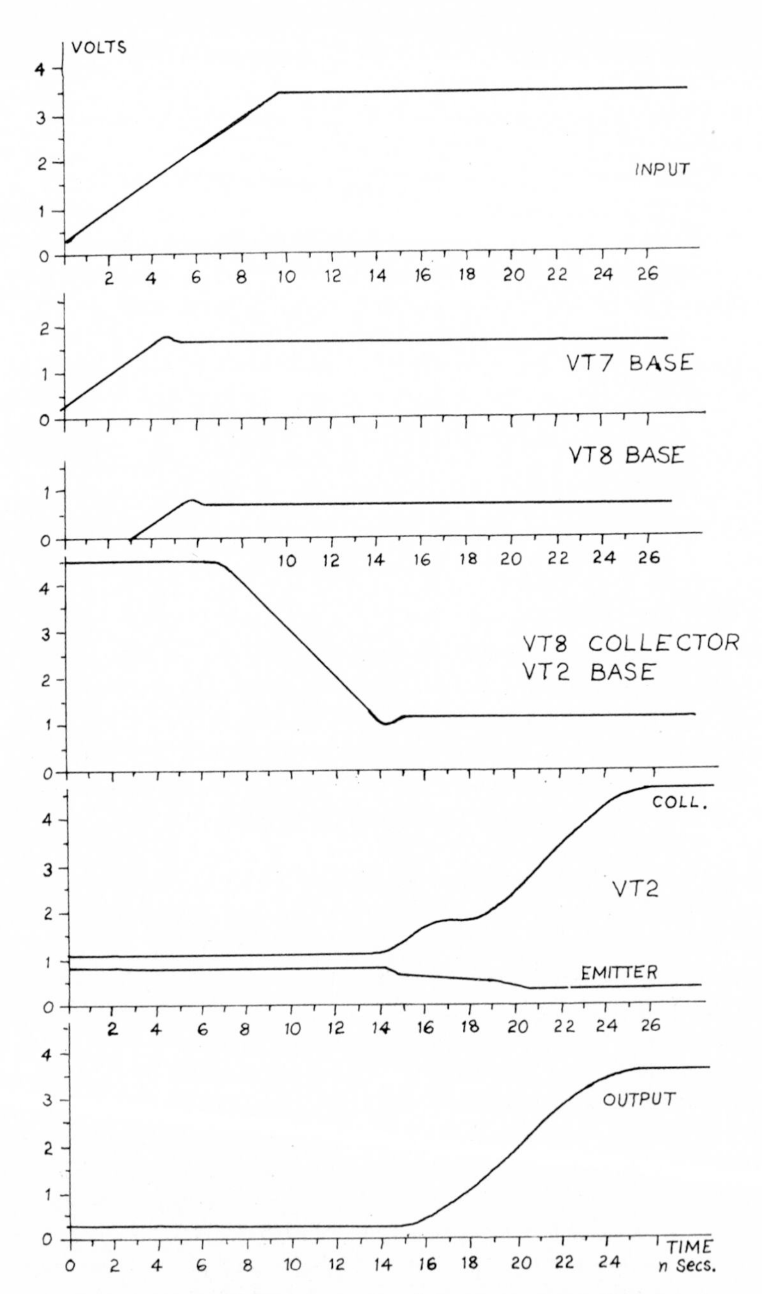

Fig. 7.2 Simplified turn-off action of non-inverting gate.

configuration VT7 and VT8. R9 and VT8 can be replaced by a diode between the emitter of VT7 and earth.

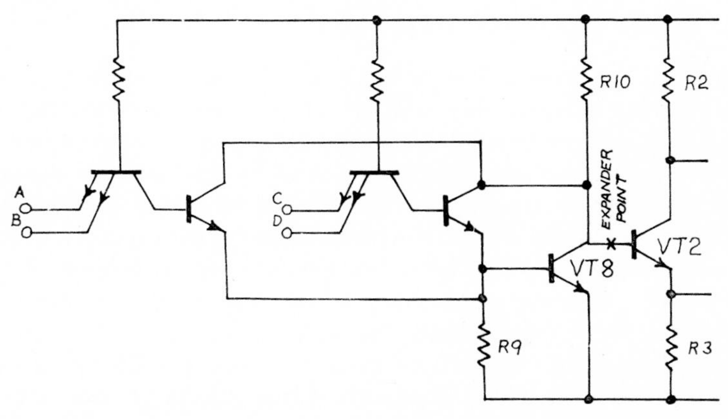

Fig. 7.3 Input stages of non-inverting gate with integral 'OR' function.

7.2 The 'NOR' Gate

The 'NOR' gate is not really a variant of the basic circuit—it is an uncommon arrangement of a normal gate. The 'NOR' gate is a basic 'AND-OR-INVERT' gate with single emitters on VT1A and VT1B. Thus it has no 'ANDing' capability, and performs only the 'OR-INVERT' or 'NOR' function. (At this point it can be noted that the basic 'NAND' gate is similarly an 'AND-OR-INVERT' gate whose 'OR' function has been reduced to unity.)

7.3 The 'OR' Gate

The 'OR' gate is an 'AND' gate (which, as explained in Section 7.1, is really an 'AND-OR-INVERT-INVERT' gate) with single emitter inputs, leaving only the 'OR-INVERT-INVERT' function.

7.4 Expanders

The use of 14-pin packages for the T.T.L. family places a restriction on the logic functions which can be performed in one package. Since power and earth take one pin each, and at least one output is necessary, not more than eleven input connections can be used. For much of the required logic this limitation is no hardship and the simpler gates are available with more than one gate per package. When large 'AND' (or 'NAND)' functions are necessary, or when a multi-way 'OR' of 'ANDed' functions is required the eleven normally available inputs may not be sufficient so arrangements can be made to use an expander package in conjunction with a basic gate package.

7.4.1 CORRECT USE OF EXPANDERS

The T.T.L. circuit is designed to optimize speed and dissipation, and to give the best possible noise rejection at its inputs. Any expansion of the logic function involves making a connection to some intermediate point in the circuit where noise immunity will be less than on the inputs. Expander packages should therefore always be placed adjacent to the packages they expand so the interconnections on the expander points are at the minimum possible length. Not only are the expander points more sensitive to noise; any capacity applied to these points will degrade the performance of the gate being expanded. (See Section 10.5.)

These degradations in performance and the susceptibility to noise of the expander points mean that expanders should always be used with care. However, provided that the limitations on the placement of expander packages are borne in mind they can be used without worry wherever they are found necessary in the logical design. When an expander package is placed adjacent to the gate it expands on any normal logic board it is quite impossible for the short tracks interconnecting the expander points to pick up sufficient noise to affect the output of the gate if the board has been designed to reasonable standards.

The real risk involved in the use of expanders lies in the temptation to use them remote from the gate being expanded. It may be tempting to ease board layout problems by using an expander in one area of the board to link (say) eight signals from that area, so that only the two expander point connections need be run right across the board to the gate where the signals are really required. This temptation must be resisted! Similarly, expanders should not be used to achieve a 'plug-in wired-OR' via the back wiring.

7.4.2 'OR' EXPANSION

'OR' expansion facilities are applied to most gates of the 'AND-OR-INVERT' type. The emitter and collector of the phase-splitter transistor VT2 are brought out to the package pins, and the expander package contains a duplicate input transistor VT1E and phase-splitter VT2E. (See Fig. 7.4.) Thus when either of the phase-splitter transistors turns on the output of the gate will go to the '0' level regardless of the state of the other phase-splitter. Fig. 7.4 shows a 'dual four' package expanded by half an expander package to achieve a three-way 'OR' function.

As more phase-splitters are connected in parallel, leakage currents will tend to turn ON the output stages, giving decreased noise immunity at the expansion points. The safe working limits are quoted by the manufacturers in their data sheets.

When it is required to expand by adding two more 'OR' functions to a gate package, the two expanders available in the same package should always be used so that the interconnecting track capacitance is minimized.

Although the 'OR' expander circuit has no earth connection, it is essential to earth the package correctly to bias the device substrate.

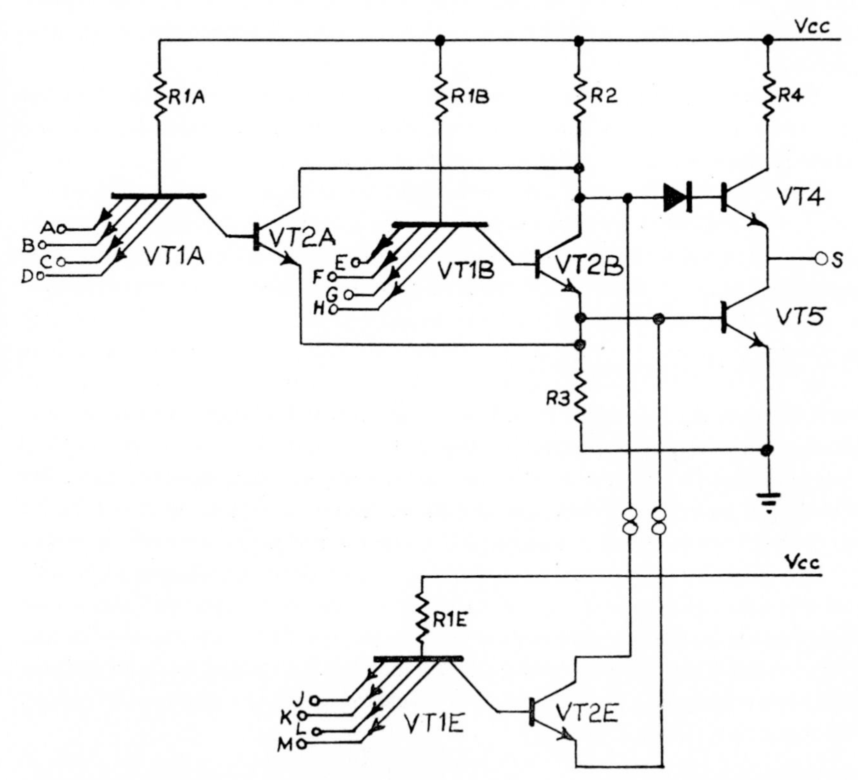

Fig. 7.4 Dual four input AND-OR-INVERT gate expanded by half a dual four-input expander to give logic function S = A.B.C.D + E.F.G.H + J.K.L.M.

7.4.3. 'AND' EXPANSION

The S.U.H.L. (and H.L.T.T.L.) families offer the facility to increase the size of a 'NAND' function beyond that offered by the normal gates. This facility is applied only to the eight-input 'NAND' gate, and it is achieved by bringing out to the package pins the base and collector of VT1. The expander is simply a multi-emitter transistor whose base and collector are commoned with those of VT1. Each expander package contains two four-emitter transistors, so the eight-input gate can be expanded to give 12 or 16 way functions with one expander package. The recommended limit is a 20 way function, i.e. one and a half expander packages.

7.4.4 NON-INVERTING 'OR' EXPANSION

The non-inverting 'OR' expansion is achieved with a single connection on the collector of VT8, so that while each input transistor is associated with a first inverter transistor VT7, each expander package also duplicates VT8.

The non-inverting expansion point is less susceptible to noise than the inverting expanders although in the (gate output) OFF (high logic level) state it does not have as much noise immunity as a normal gate input. Since VT2 switches only near one end of the swing on the collector of VT7, the effects of leakage currents rising as expanders are added can be ignored provided that expanders are used only adjacent to the gate package being expanded, which automatically limits the number of expanders.

7.5 Line Drivers or Buffers

A line driver is similar to a basic gate, but with larger transistors and lower value resistors to give increased drive capability for driving 100 Ω lines. Series 74 buffers have R1 the same value as in a standard gate, but S.U.H.L. and 9000 buffers have a lower value for R1 so that the buffer or line driver presents a loading of two gates to the gate which drives it.

If a package containing four two-input line drivers is being used with all outputs fully loaded, the level of an ON (0 level) gate will change as other gates in the package are turned ON due to the extra current flowing the common earth return path. All devices are d.c. tested with all outputs fully loaded so the change should not cause the already ON output to exceed the specified '0' level voltage.

It should be noted that the quoted device dissipations do not include any allowance for dissipation due to load currents in the output transistors and the common earth path, so that when chip temperatures are being calculated this dissipation should be added to the quoted figure.

The terms 'line driver' and 'buffer' have been treated as being synonymous in this section because some manufacturers use one name and some use the other. The use of the name 'line driver' does not imply that the devices are capable of driving into a single d.c. load to earth or the supply rail of 100 Ω or thereabouts (see Section 17.3.1.).

7.6 Lamp Drivers

The T.T.L. lamp driver is made from a line driver chip with the metallization to VT3 and VT4 omitted so the collector of VT5 is connected to the output pin only. (See Fig. 7.5.) The output current which can be sunk in the '0' state is usually increased, but the '0' level voltage is raised to allow for this increased current.

Whenever a lamp driver is used to drive normal T.T.L. gates, a 'pull-up' resistor to the 5 V rail should be provided. The value of the resistor can be calculated to suit the loading on the gate. The absence of the active

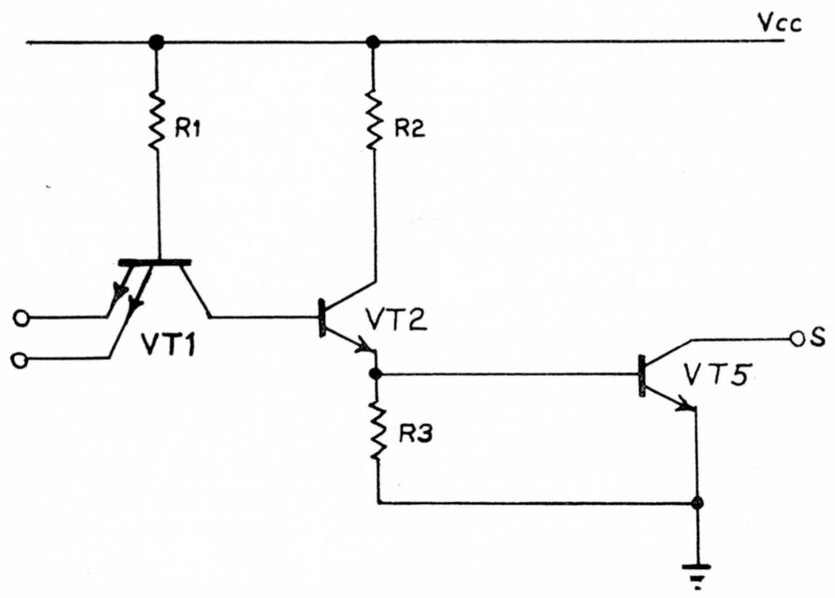

Fig. 7.5 T.T.L. Lamp driver or open-collector gate.

pull-up circuit means that lamp drivers can have their outputs interconnected to give a 'wired-OR' function (but see Section 14.1.4.6).

Lamp drivers can also be used as delay generators by connecting a capacitor from the output to earth to slow down the rising edge, and using another T.T.L. gate or lamp driver to 'square-up' the output waveform. At delays of more than about one microsecond, the rising edge into the second gate becomes so slow that the edge speeds of the second gate are seriously degraded and the action becomes unreliable. Such delay circuits do not tolerance at all well, but where the delay need not be precise or where the use of a variable resistor can be tolerated they are very convenient.

As with the line driver, device dissipation with all outputs fully loaded and ON simultaneously can be very high. Significant changes in '0' level voltages can be measured as other gates in the same package are switched on or off, although all outputs should at all times remain below the specified '0' level. It is recommended that when lamp drivers are being used to drive full loads, arrangements should be made to ensure that not more than two gates in any one package can be ON simultaneously. If three or more gates in any one package are to be on simultaneously, their load currents should be restricted. A safe working level for long-term reliability should result if $0{\cdot}06_n + 10(I_1^2 + I_2^2 + I_3^2 + I_4^2)$ is less than 0·300 when n is the number of gates ON simultaneously and I is in amperes.

8

Range of Gates within the Families

8.1 Nomenclature

8.1.1 INTRODUCTION

The total range of T.T.L. gates available is very large indeed. One manufacturer lists over one hundred gate types in a publicity booklet, and there are more types made by other manufacturers which are not listed in this booklet.

The existence of military and commercial ranges, and high and low fan-outs in the S.U.H.L. families, makes the list of type numbers a long one.

Such a wide choice of gate types can make standardization and spares holding quite difficult, and although the wide range of gate types makes it easy for a designer to select devices to implement his own particular logic design, it also makes it harder for him to remember what is available.

8.1.2 MANUFACTURERS' IDENTIFICATION

8.1.2.1 *Manufacturers' names*

Discussions between designers working with different T.T.L. families can be hindered by a lack of standardization in the naming of the gates. The name 'quad two-input NAND gate' is fairly simple (or does it mean four NAND gates with the outputs 'ORed' together?), but 'majority decision gate' is not very explicit, and the 'exclusive-OR' (which does not perform the exclusive-OR of two positive logic functions as its name might suggest) is somewhat misleading. There has been a tendency to shorten 'AND-OR-INVERT' to 'OR', but since direct 'OR' gates are available this can be dangerous.

8.1.2.2 *Type numbers*

If the names of the gates can not be used, then perhaps the manufacturers' type numbers could be learnt. Such an exercise would be all right for designers whose hobby is memorizing railway timetables or telephone directories, but as a practical means of overcoming the confusion between manufacturers the use of type numbers is worse than the use of names. Each of the major families has its own basic numbering system, and this system is generally used by 'second-source' manufacturers, but there are other manufacturers who have their own numbering systems. Some of the systems have some obvious logic in the choice of numbering, but others appear to be quite arbitrary. Even in the families where some logical system can be determined, the 'system' usually breaks down. For instance, a 9002 gate has two inputs, a 9003 has three inputs, a 9004 has four inputs, but the eight-input gate is numbered 9007! Some manufacturers vary the last digit of the type number to indicate military or commercial devices. H.L.T.T.L. has an odd numbered last digit for military, but S.U.H.L. has an even last digit! M.T.T.L. varies the first digit, and Series 74 has the Series 54 as its military equivalent. It would require a prodigious feat of memory to memorize all type numbers.

As an example, a 9002 gate could be replaced by a 7400 or an MC 3000 (with differences in speed), and, if a board has not been laid out, by an SG 143, TNG 3414, MC 458, SG 223, MC 1051, or TNG 3444. And then there are the military versions of the same device which give a further eight numbers such as MC 558, 5400, TNG 3443, and also high fan-out versions in seven of the families listed (such as TNG 3412, MC 408). Add variations in prefix letters (such as RG 223) for other manufacturers who follow one of the original manufacturer's numbering schemes, and it is obvious that the use of manufacturers' type numbers will not avoid confusion when engineers using different families meet to discuss design problems.

8.1.3 SIMPLIFIED NUMBERING SYSTEM

In an endeavour to eliminate most of this confusion, the author evolved a logical numbering system which enables designers to work out the number for any gate or package type, and which can be translated to any manufacturer's type number for ordering purposes. This system was used with considerable success within Computer Division, the Marconi Company, and was also used in correspondence with manufacturers of T.T.L.

8.1.3.1 *Basic numbering system*

The basic system uses three digits. The first digit is the number of logically independent circuits in the package; the second digit is the minimum number of 'ANDed' inputs, and the third digit is the number of phase-splitter transistors 'ORed' together, or with an 'OR' expansion point available.

Thus a 'quad two-input NAND gate' is simply described as a '420', and cannot be confused with a 'quad two-input AND-OR-INVERT' gate (124) or a 'quad two-input NOR' gate (412). However, it could be confused with a 'quad two-input AND' (or non-inverting) gate. Also, a 'dual four-input NAND' gate can not be distinguished from a 'dual four-input buffer' (which performs the 'NAND' function). This led to the introduction of a fourth symbol in the form of a letter with a clear significance. Further possible confusions led to the addition of a fifth symbol, also a letter with an obvious meaning.

The fourth symbols cover S for slow, F for fast, B for buffers, N for non-inverting, L for lamp driver, T for a gate with transient control, C for a gate with complementary outputs, and E for an 'AND'-expandable 'NAND' gate.

Fifth symbols describe the gate's output specification, and are L for low fan-out, H for high fan-out, and E for expanders.

8.1.3.2 *Flip-flops and complex functions*

This simple code can not readily be applied to flip-flops and M.S.I. functions, but these devices can be given five character codes which define their functions. For flip-flops, the first digit defines the number of circuits in the package, the second symbol generally describes the input gating, and the third and fourth symbols define the type of flip-flop. Thus 1AJK is a J.K. flip-flop with 'ANDed' input gating; 1OJK is a J.K. with 'ORed' inputs; 21DF is a single-input dual D type; IJK indicates a flip-flop with inverted or complemented ($\bar{J}$ or $\bar{K}$) inputs, and EJK or EMS indicates edge-triggered devices. The fifth character indicates the speed range.

M.S.I. elements use similar coding, with the first digit indicating the number of bits. The second symbol defines the general types of device, and the third and fourth symbols define more closely the working of the device. Thus 4A indicates a four bit adder, 4ASC is a serial (or ripple) carry four bit adder, and 4APC is a parallel carry (or 'look-ahead' carry) adder. 8RSP is an eight bit register with parallel outputs, and 4RSS is a four bit shift register with serial output. For counters, the last two characters define the number counted, e.g., 4C16, or 4C12, etc.

There is far less need for a 'universal numbering scheme' for M.S.I. elements than there is for gates because there is not the same diversity of type numbers covering the same or similar devices, and the numbering scheme is useful mainly for scheduling purposes.

8.1.3.3 *Prefix and suffix*

Five-symbol type numbers as described above are generally all that a logic or board designer needs. There could still be confusion in ordering between military and commercial ranges, and possibly between S.U.H.L., 9000, and Series 74 pin configurations. These can be resolved by a suffix and a prefix. The suffix is M or C (for Military or Commercial), and the prefix is S for S.U.H.L. types, 7 for Series 74, and 9 for Series 9000.

8.1.3.4 *Application to computer-aided design*

One advantage of this simplified numbering system is that it can be used directly in computer programmes when computer-aided design is used to check logical circuits and to design printed circuit boards. If it is used with no stored 'translation schedule', there will be some loss of logical efficiency in the use of gates, because several of the more complex gates do not have the same number of 'ANDed' inputs to each part of the gate (see Section 8.2.2) and the computer might allow only the number listed for all parts of the device.

Where computer-aided design is used, 'imaginary' device type numbers can be used, such as 190 or 144, and it can be left to the computer to determine what is in a gate package and whether expanders are required. The computer can also determine whether high or low fan-out devices should be used, and possibly even the speed range required.

8.2 Gate Types

8.2.1 'NAND' GATES

The two-input NAND gate (see Fig. 8.1) (generally used throughout this book as the basic T.T.L. gate) is the most widely used gate type, and all manufacturers of T.T.L. include this gate in their range. Each gate requires three package pins, so four gates (plus power and earth pins) use all 14 pins of a standard D.I.P. or Flat-Pack. This gives the 'quad two-input NAND' gate package, or 420. All the major families also include triple three-input, dual four-input, and single eight-input NAND gates (330, 240, 180).

The 'hex-inverter' (610) which is available from most, if not all, manufacturers, can also be included with the 'NAND' gates, although it performs no logic function.

The only variant which is not offered by all manufacturers is an 'AND'-expandable eight-input 'NAND' gate. The ability to increase the size of a single 'NAND' function to 12, 16, or even more ways can be very valuable since the alternative, when the expansion facility is not available, is to group the inputs with non-inverting 'AND' gates to reduce the size of the 'NAND' function to eight ways. If 'AND' gates are not available then the only alternative to an expanded 180 is to perform the logic in three cascaded stages, i.e. a pair of 180s, each feeding inverters, and the inverted outputs in turn feeding into a further 'NAND' gate. This can prove expensive in packages and in propagation time.

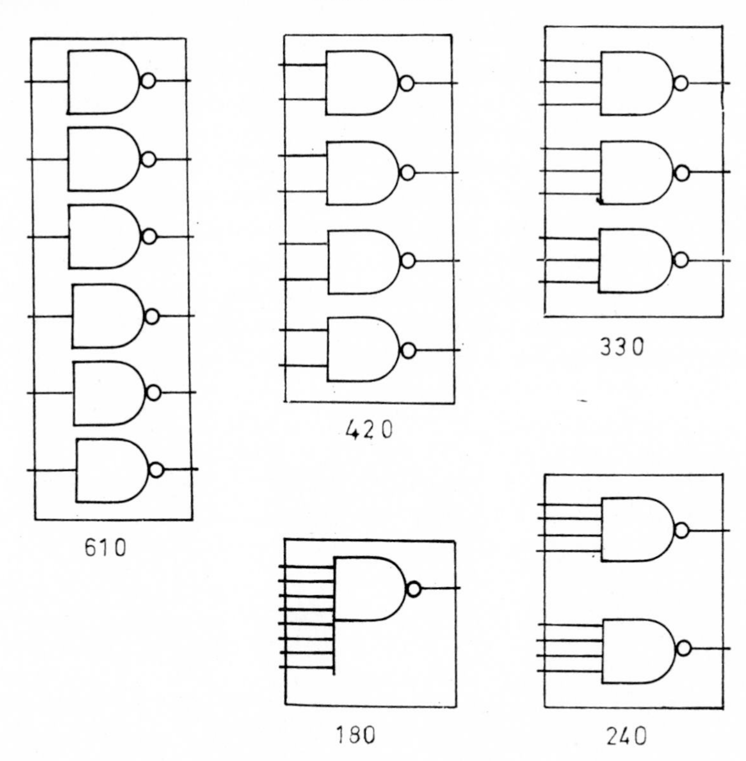

Fig. 8.1 NAND gate types.

8.2.2 'AND-OR-INVERT' GATES

The 'pull-up' network of VT3 (or D3) and VT4 in a normal T.T.L. gate precludes the use of the 'wired-OR'—the achieving of a logical 'OR' function by direct interconnection of gate outputs. This means that to keep down the numbers of packages used and to keep propagation delays down to reasonable limits the manufacturers of T.T.L. had to provide gates which include an internal 'OR' facility (see Section 3.1.2).

It is in the selection of 'AND-OR-INVERT' types that the different families of T.T.L. first show a difference which can affect the would-be user. So far as is known to the author at the time of writing there is no 'AND-OR-INVERT' type common to all families. The nearest is the 'dual dual-two-input-gate' (222). This basic gate type requires twelve package pins (eight for inputs, two for outputs and power and earth). It is in the allocation of the remaining two pins that Transitron's H.L.T.T.L. 1 and 2, and Raytheon's RAY 2 differ from other manufacturers. They use the two 'spare' pins for a third input to one of the 'AND' inputs in each half of the gate, to give a logic function of $S = A.B + C.D.E$ (twice in the package). Other manufacturers use the 'spare' pins for 'OR' expansion points on one of the two gates (see Fig. 8.2) or leave them unused.

The next nearest to a 'common' device is the 'quad two-input' (124), which is available in all families except M.T.T.L. 3. A basic 124 requires only eleven pins (eight inputs, one output, power and earth) and most manufacturers use two of the 'spare' pins as 'OR' expansion points and the third as an extra 'AND' input so the logic function of the package is $\overline{S} = A.B + C.D + E.F + G.H.J +$ (see Fig. 8.2).

The Series 54/74 version of the 124 has two inputs to each 'AND' function and leaves one pin spare. Some firms offer 222 and 124 types with and without the 'OR' expansion facility.

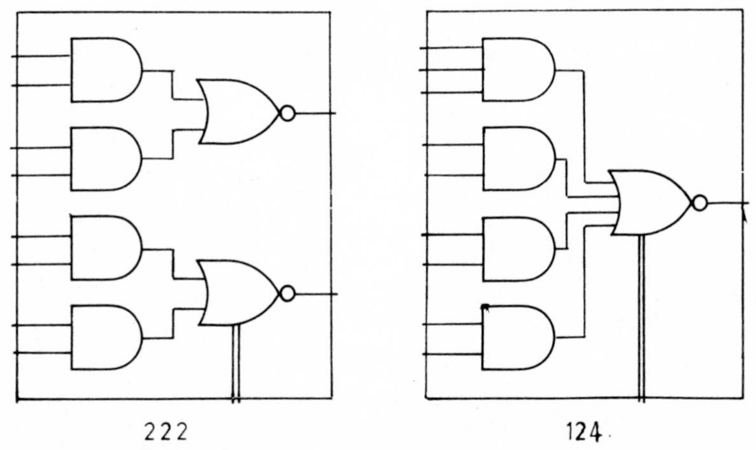

Fig. 8.2 Common AND-OR-INVERT gate types.

The 'main building bricks' of any T.T.L. system (almost regardless of family) are the 610, 420, 330, 240, 180, 222, and 124. All other types mentioned in this chapter are 'specials', available in only some of the families. Generally, it is the S.U.H.L., H.L.T.T.L., and M.T.T.L. families which offer the widest variety of gate types.

Other 'AND-OR-INVERT' types available in some families only are the 'triple three-input', and 'dual four-input' gates (133 and 142), both of which are usually 'OR'-expandable.

An 'exclusive-OR with complement' is also available. The main logic function is an 'OR' of two three-input 'AND' gates, with inverted output, but this inverted output is internally fed to a further three-input 'NAND' gate from which the complemented output is taken. Under the simplified numbering system this gate is a 132C. (See Fig. 8.3.)

A package which can be considered with the 'AND-OR-INVERT' types is the quad two-input 'NOR' gate (412) available from some manufacturers. This is an 'AND-OR-INVERT' type with only one emitter used on VT1 in each circuit.

When 'AND-OR-INVERT' gates are used, it may not be possible or necessary to use all the 'AND' input gates in the package. A typical case arises in high speed arithmetic units, where an 'OR' of 1, 2, 3, and 4 way 'AND' functions

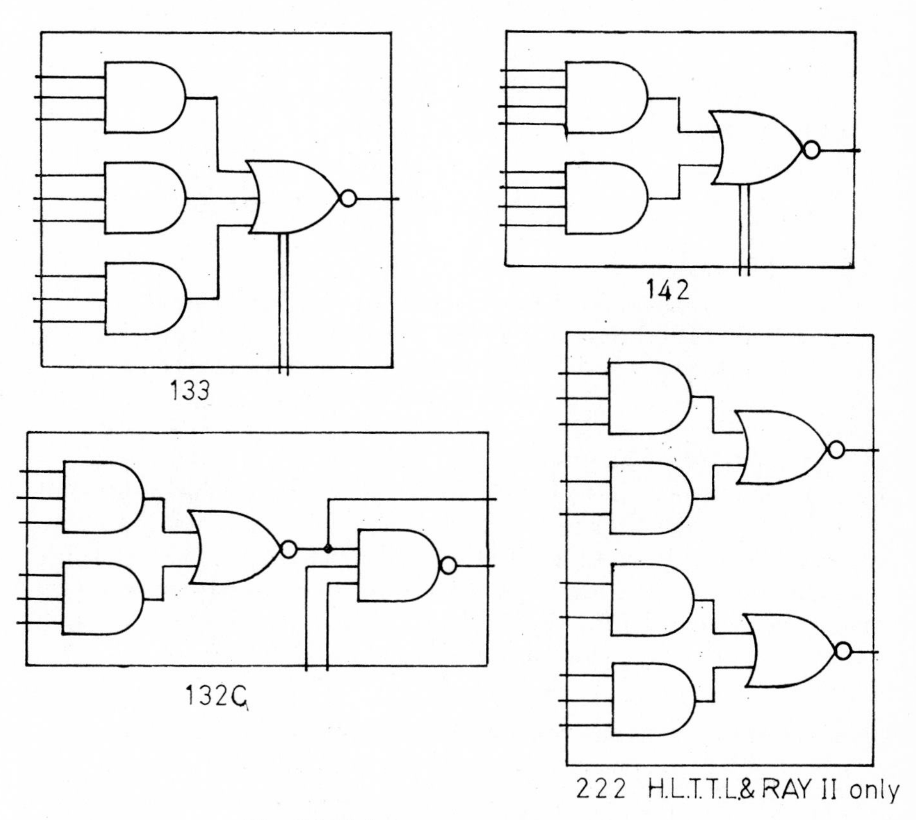

Fig. 8.3 Other AND-OR-INVERT gate types.

is required. This can be realized with a 124 package (with one three-input gate) plus half a 241 expander package. Thus one 'AND' gate in the 124 package will have its inputs strapped together for the single way function, and one 'AND' gate will be unrequired. At least one input to this unused gate must be earthed. If the unused gate inputs are left floating, the resulting '1' output from the 'AND' gate will inhibit the 'OR' function. This is a very obvious point, but it has probably been the cause of more boards having to be re-laid than any other single error. (See Section 9.5.4.)

8.2.3 'AND' OR NON-INVERTING GATES

Non-inverting gates are available from only a few of the manufacturers of T.T.L. The commonest gate package is a dual four-input 'AND' gate with an 'OR' expansion facility (241N). Other types available include a quad two-input (420N), dual four-input with transient control (240T), dual three-input 'OR'-expandable with transient control (231T), dual dual two-input 'OR'-expandable (222N) and a quad two-input 'OR' gate (412N).

8.2.4 EXPANDERS

8.2.4.1. *'And' expanders*

There is only one 'AND' expander in T.T.L., a dual four-input device (240*x*E), for use with the expandable eight-input 'NAND' gates.

8.2.4.2 *'Or' expanders*

Two 'OR' expanders for use with 'AND-OR-INVERT' gates are featured in several of the T.T.L. ranges; a dual four-input with separate outputs (241*x*E), and a quad two-input with a single logical output (124*x*E). This latter expander has two two-input and two three-input 'AND' gates.

8.2.4.3 *Non-inverting expanders*

The most common non-inverting 'OR' expander is a 'dual dual-two-input' type (222NE) which has a two- and a three-input 'AND' gate 'ORed' together in each circuit. Other types available are triple three-input (133NE) which has one three- and two four-input 'AND' gates 'ORed' to the single output, and an '8 + 3 input' (132NE) which, as its name suggests, has an eight-input 'AND' gate 'ORED' with a three-input gate.

8.2.5 BUFFERS, LAMP DRIVERS, AND OTHER TYPES

The most common T.T.L. buffer is the dual four-input 'NAND' type (240B), but quad two-input versions are available from some manufacturers. Similar lamp drivers are available from a few manufacturers only.

Some manufacturers offer a quad two-input 'NAND' gate without the output pull-up circuit, so that 'wired-OR' functions can be implemented.

Other devices such as 'pulse shapers' or 'delay generators' are made by several manufacturers and are included in their lists of gates available. These are regarded as special circuits, and are not discussed in any detail in this book.

9

Main Parameters of T.T.L. Gates

9.1.1 Introduction

The main parameters of any logic which affect the circuits which can be designed with the logic are the speed, fan-out, noise immunity, and dissipation.

With T.T.L. these are all inter-related. Speed is affected by the gate connections; noise immunity is based on arbitrarily chosen voltage levels to give a nominal voltage noise margin, and the fan-out is determined with respect to these arbitrarily chosen voltage levels.

Changes in supply voltage, temperature, etc. can affect these parameters. Since again the effects of changes in working conditions are all inter-related, this chapter discusses the basic parameters under normal conditions, and Chapter 10 deals with the various degradations due to changes in working conditions. Sections in Chapter 10 are numbered with reference to Chapter 9; i.e. Sections 10. ?.2 deal with degradations in speed. Section 10.4.3 does not exist (degradation in fan-out with fan-out), and Section 10.4.4 (degradation in noise immunity with fan-out) is covered in Section 9.3 on fan-out.

'Rules' can be drawn up to guarantee to keep all designs within the chosen noise margin limits, to restrict the design of printed circuit boards such that cross-talk and ring cannot cause spurious switching, and to enable the gates to run at their maximum possible speed. Such worst-case rules can prove crippling to practical economic design, and the designer must judge for himself whether to abide by the rules, possibly at the expense of economy, or whether to allow a slight relaxation on speed or noise margin to achieve a required fan-out. Each section of this chapter discusses first the specified limits on the parameters; then consideration is given to the typical figures found when devices are tested. Designers must be prepared for disappointments if they attempt to work to the typical figures, but the figures can be used as a guide when the worst case rules are found to be excessively restrictive.

TABLE 9.1 Main parameters

		Output					
		Low		High			
Temp. Range °C	Manufacturer Family	V_{OL}	I_{OL} mA	V_{OH}	$-I_{OH}$ mA	I_{OLK} mA	$-I_{OS}$ mA
−55 to +125	Sylvania. S.U.H.L. 1	**0·40** (0·40–0·45)	**10** **or** **20**	**3·2** (2·8–3·35)	**0·7** **or** **1·5**	**0·250**	**10** **to** **45**
	Raytheon RAY 1	0·40 (0·40–0·45)	10 or 20	3·2 (2·8–3·35)	0·7 or 1·5	0·250	10 to 45
	Motorola M.T.T.L. 1	0·40 (0·40–0·45)	10 or 20	**2·4** (2·5–2·7 at $V_{CC\ min}$)	0·7 or 1·5	0·250	10 to 45
	Transitron H.L.T.T.L. 1	0·45	10 or 20	**3·2** (2·8–3·35)	0·7 or 1·5	0·250	10 to 45
0 to +75	Sylvania S.U.H.L. 1	**0·40** (0·40–0·45)	10 or 20	**3·1** (3·0–3·15)	**0·6** **or** **1·2**	0·250	**10** **to** **60**
	Raytheon RAY 1	0·40 (0·40–0·45)	10 or 20	3·1 (3·0–3·15)	0·6 or 1·2	0·250	**10** **to** **45**
	Motorola M.T.T.L. 1	0·40 (0·40–0·45)	10 or 20	**2·4** (2·5–2·5 at $V_{CC\ min}$)	0·6 or 1·2	0·250	10 to 45
	Transitron H.L.T.T.L. 1	**0·45**	10 or 20	**3·1** (3·0–3·15)	**0·7** **or** **1·5**	0·250	10 to 45
0 to 70 and −55 to +125	Texas In. 54/74	**0·4**	16	**2·4** at $V_{cc\ min}$	**0·4**	—	**18 to** **55** 20–55 on ser. 54
	Sylvania and Transitron 54/74	0·4	16	2·4 at $V_{CC\ min}$	0·4	—	18 (20) to 55
	National 54/74	0·4	16	2·4 at $V_{CC\ min}$	0·4	—	18 (20) to 55
0 to +70	Texas In. Series 10	**0·5**	**13**	**2·4** at $V_{CC\ min}$.	**0·8**	—	**12** **to** **55**
	Texas In. System 11	**0·35**	**16**	2·4 at $V_{CC\ min}$.	0·8	**1·0**	**18** **to** **55**

specified by manufacturers

Input				General							
Low		High		Speed			Edges		Power		Fan-out
$-I_{IL}$ mA	$V_{IL\ max.}$	I_{IH} mA	$V_{IH\ min.}$	t_{pd} ns	t_{PHL} ns	t_{PLH} ns	t_{THL} ns	t_{TLH} ns	I_{CCL} mA	I_{CCH} mA	
1·33	**1·2** (1·0–0·9)	**0·1** at 4·5V	**1·7** (2·0–1·4)	—	**20**	**20**	**5**	**8**	**6**	**3**	**7 or 15**
1·33	**1·2** (1·0–0·85)	0·1	1·7 (2·0–1·4)	—	20	20	5	8	6	3	7 or 15
1·33	**1·2** (1·0–0·9)	0·1	1·7 (2·0–1·4)	—	20	20	5	8	6	3	7 or 15
1·33	1·2 (1·0–0·9)	0·1	**1·8** (2·0–1·4)	**18**	—	—	—	—	6	3	7 or 15
1·66	**1·2** (1·1–1·1)	0·1	**1·8** (1·9–1·7)	—	**20**	**20**	**5**	**8**	**7·5**	3	**6 or 12**
1·66	1·2 (1·1–1·1)	0·1	1·8 (1·9–1·7)	—	20	20	5	8	7·5	3	6 or 12
1·66	1·2 (1·1–1·1)	0·1	1·8 (1·9–1·7)	—	20	20	5	8	7·5	3	6 or 12
1·33	1·2 (1·1–1·1)	0·1	1·8 (1·9–1·7)	**18**	—	—	—	—	**6**	3	**7 or 15**
1·6	**0·8**	**0·04** at 2·4V	**2·0**	—	**15**	22	—	—	**5·5**	**1·3**	**10**
1·6	0·8	0·04	2·0	—	15	22	—	—	5·5	**2**	10
1·6	0·8	0·04	2·0	—	15	25	—	—	**5·1**	**1·8**	10
1·6	—	**0·1**	—	—	**25** at 50 pF	**50** at 50 pF	—	—	**3 typ.**	**1 typ.**	8
1·6	**0·95**	**0·04** at 2·4V	**1·7**	—	**10 3 min.**	**20 6 min.**	—	—	**6**	**2**	**10**

TABLE 9.1 Main parameters

		Output					
		Low		High			
Temp. Range °C	Manufacturer Family	V_{OL}	I_{OL} mA	V_{OH}	$-I_{OH}$ mA	I_{OLK} mA	$-I_{OS}$ mA
−55 to +125	9000	**0·4**	**17·6** at $V_{CC\,max.}$ 13·6 at $V_{cc\,min}$	2·4 at $V_{CC\,min.}$	**1·32**	—	—
0 to +70	9000	**0·45**	**16·0** at $V_{CC\,max.}$ 14·1 at $V_{cc\,min}$	2·4 at $V_{CC\,min.}$	**1·2**	—	—
−55 to +125	Sylvania S.U.H.L. 2	**0·40** (0·40–0·45)	12 **or** **22**	**3·1** (2·7–3·15)	**1·2** **or** **2·2**	**0·250**	**25** **to** **100**
	Raytheon Ray 2	0·40 (0·40–0·45)	12 or 22	3·1 (2·7–3·15)	1·2 or 2·2	0·250	25 to 100
	Motorola M.T.T.L. 2	0·40 (0·40–0·45)	12 or 22	**2·4** (2·5–2·5 at $V_{CC\,min.}$)	1·2 or 2·2	0·250	25 to 100
	Transitron H.L.T.T.L. 2	**0·45**	**11** **or** **22**	**3·2** (2·8–3·35)	**0·75** **or** **1·5**	0·250	**30** **to** **80**
0 to +75	Sylvania S.U.H.L. 2	**0·40** (0·40–0·45)	**12·5** **or** **22·5**	**3·0** (2·9–3·0)	**1·0** **or** **1·8**	0·250	**25** **to** **100**
	Motorola M.T.T.L. 2	0·40 (0·40–0·45)	12·5 or 22·5	**2·4** (2·5–2·5 at $V_{CC\,min.}$)	1·0 or 1·8	0·250	25 to 100
	Transitron H.L.T.T.L. 2	**0·45**	**11** **or** **22**	**3·1** (3·0–3·15)	**0·75** **or** **1·5**	0·250	**30** **to** **80**
0 to +70 and −55 to +125	Texas In. 54H 74H	**0·4**	**20**	**2·4** at $V_{CC\,min.}$	**0·5**	—	**40** **to** **100**
−55 to +125	Raytheon RAY 3	0·40 (0·40–0·45)	**22**	**3·1** (2·7–3·15)	**2·2**	**0·250**	**25** **to** **100**
0 to 70	Raytheon RAY 3	0·40 (0·40–0·45)	**22·5**	**3·0** (2·9–3·0)	**1·8**	0·250	25 to 100

cified by manufacturers—*contd.*

Input				General							
Low		High		Speed			Edges		Power		Fan-out
$-I_{IL}$ mA	$V_{IL\ max.}$	I_{IH} mA	$V_{IH\ min.}$	t_{pd} ns	t_{PHL} ns	t_{PLH} ns	t_{THL} ns	t_{TLH} ns	I_{CCL} mA	I_{CCH} mA	
·6	**0·9** (0·8–0·8)	**0·06** at 4·5V	1·7 (2·0–1·4)	—	**12 3 min.**	**10 3 min.**	—	—	**5·5**	**1·6**	**11**
·6	**0·85**	0·06 at 4·5V	**1·8** (1·9–1·6)	—	**15 3 min.**	**13 3 min.**	—	—	**6·1**	**1·7**	**10**
·0	**1·1** (0·9–0·9)	**0·1** at 4·5V	1·8 (2·0–1·6)	—	**10**	**10**	**2·5**	**4·0**	**7·5**	**3·5**	**6 or 11**
·0	1·1 (0·9–0·85)	0·1 at 4·5V	**1·7** (2·0–1·4)	—	10	10	2·5	4·0	7·5	**3·75**	6 **or** 11
·0	1·1 (0·9–0·9)	0·1 at 4·5V	1·7 (2·0–1·4)	—	10	10	2·5	4·0	7·5	3·75	6 or 11
·2	**1·2** (1·0–0·9)	**0·15**	1·7 (2·0–1·4)	—	10	10	—	—	**8**	**4**	**5 or 10**
·5	**1·1** (1·0–1·0)	**0·1** at 4·5V	**1·8** (1·9–1·7)	—	10	10	**2·5**	**4·0**	**10**	**5**	**5 or 9**
·5	1·1 (1·0–1·0)	0·1 at 4·5V	1·8 (1·9–1·7)	—	10	10	2·5	4·0	10	5	5 **or** 9
·2	**1·2** (1·1–1·1)	**0·15**	1·8 (1·9–1·7)	—	10	10	—	—	**8**	**4**	**5 or 10**
·0	**0·8**	**0·05**	**2·0**	—	**10**	**10**	—	—	**10**	**4·2**	**10**
·0	**1·2** (1·2–0·9)	0·05	**1·7** (2·0–1·4)	—	**6**	**5**	**2·5**	**3·0**	**7·5**	**3·5**	**11**
·5	1·2 (1·1–1·0)	0·05	**1·8** (1·9–1·7)	—	6	5	2·5	3·0	**10**	5	**9**

TABLE 9.1 Main paramete

		Output					
		Low		High			
Temp. Range °C	Manufacturer Family	V_{OL}	I_{OL} mA	V_{OH}	$-I_{OH}$ mA	I_{OLK} mA	$-I_{OS}$ mA
0 to +70 and −55 to +125	Motorola M.T.T.L. 3	**0·4**	**23 at** $V_{CC\ max}$. **19 at** $V_{CC\ min}$.	**2·5** at $V_{CC\ min}$.	**2**	—	**30 to 100**
	Texas In. 54L 74L	**0·3**	**2**	**2·4** at $V_{CC\ min}$.	**0·1**	—	**3 to 15**
	Texas In. 54S 74S	**0·5**	**20**	**2·7** at $V_{CC\ min}$.	**1·0**	—	**40 to 100**
	Buffers						
−55 to +125	Sylvania S.U.H.L. 1	**0·40** (0·40–0·45)	**20 or 40**	**3·2** (2·8–3·35)	**1·5 or 3·0**	**0·250**	**50 to 150**
	Transitron H.L.T.T.L. 1	**0·45**	**32 or 48**	3·2 (2·8–3·35)	**2·4 or 3·6**	0·250	**50** to **150**
0 to +70	Sylvania S.U.H.L. 1	**0·40** (0·40–0·45)	**20 or 40**	3·2 (2·8–3·35)	**1·2 or 2·4**	0·250	50 to 150
	Transitron H.L.T.T.L. 1	**0·45**	**32 or 48**	3·2 (2·8–3·35)	**2·4 or 3·6**	0·250	50 to 150
−55 to +125	9000	**0·4**	**52·8 at** $V_{CC\ max}$. **40·8 at** $V_{CC\ min}$.	**2·4** at $V_{CC\ min}$.	**3·96**	—	—
0 to +70	90C0	**0·45**	**48 at** $V_{CC\ max}$. **42·3 at** $V_{CC\ min}$.	2·4 at $V_{CC\ min}$.	**3·6**	—	—
0 to +70 and −55 to +125	54 74	**0·4**	**48**	2·4 at $V_{CC\ min}$.	**1·2**	—	**18 to 70** 20–7(on Ser.

specified by manufacturers—*contd.*

Input				General							
Low		High		Speed			Edges		Power		Fan-out
$-I_{IL}$ mA	$V_{IL\ max.}$	I_{IH} mA	$V_{IH\ min.}$	t_{pd} ns	t_{PHL} ns	t_{PLH} ns	t_{THL} ns	t_{TLH} ns	I_{CCL} mA	I_{CCH} mA	
2·3	**1·1** (1·1–0·9)	**0·08**	**1·8** (2·0–1·8)	—	**10**	**10**	—	—	**9**	**4·3**	**10**
0·18	**0·7**	**0·01** at 2·4V	**2·0**	—	**60**	**60**	—	—	**0·46**	**0·18**	**10**
2·0	**0·8**	**0·10**	**2·0**	—	**5**	**4·5**	—	—	**9**	**4·25**	**10**
2·0	**1·2** (1·0–0·9)	**0·2**	**1·7** (2·0–1·4)	—	**25 into 150pF**	**25 into 150pF**			**14**	**4·5**	**15 or 30**
2·5	1·2 (1·0–0·9)	0·2	**1·8** (2·0–1·5)	**18 into 100pF**	—	—			**12**	**6**	
2·5	**1·2** (1·1–1·1)	0·2	**1·8** (1·9–1·7)	—	**25 into 150pF**	**25 into 150pF**			**17**	**5·5**	**15 or 30**
2·5	1·2 (1·1–1·1)	0·2	1·8 (1·9–1·7)	**18 into 100pF**	—	—			**12**	**6**	
3·2 at V_{CC} max. 2·82 at V_{CC} min.	**0·9**	**0·12**	**1·7** (2·0–1·4)	—	**10**	**15**			**12·9**	**3·2**	**33**
3·2 at V_{CC} max. 2·48 at V_{CC} min.	**0·85**	0·12	**1·8** (1·9–1·6)	—	**13**	**17**			**14·6**	**3·4**	**30**
1·6	**0·8**	**0·04 at** 2·4V	**2·0**	—	**15**	**29**			**8·6**	**2**	**30**

9.1.2 Manufacturers' Specifications

Table 9.1 gives an abridged summary of the main parameters specified by the major manufacturers of T.T.L. The data sheets supplied by the manufacturers vary widely. There is some confusion over the naming of the various parameters, and test levels for the various parameters vary from manufacturer to manufacturer.

S.U.H.L. devices are generally specified with V_{CC} at 5 V whereas Series 54/74 devices are specified with the lowest or highest permissible rail voltage, the limit for each parameter being that which results in the worst-case reading. Series 54/74 devices are usually specified at the limit figure for the entire temperature range, but S.U.H.L. specifications usually have three columns, one for each of the limit temperatures and one for 25°C. Series 9000 specifications quote figures for three temperatures, and at worst case V_{CC}.

Generally, manufacturers seem to copy the data sheets of the firm which first made the logic family, this being most obvious in the case of Transitron Electronic, whose H.L.T.T.L. 4 (Series 74) specifications follow the style adopted by Texas Instruments instead of continuing the style they used for their H.L.T.T.L. 1 and 2 (S.U.H.L. type). However, there are differences between manufacturers' versions of the same family, and it should never be assumed that all parameters will be exactly the same.

It should be noted that the figures quoted in Table 9.1 do not always apply to all the gates in a family. To keep as fair a comparison as possible, all information in the table has been taken from the data sheets normally supplied by sales representatives, and all the gate data were taken from the sheets applying to the quad two-input 'NAND' gates. Data for buffers and line drivers were taken generally from the dual four-input devices. In all cases the data apply to one gate only in the package (i.e. I_{CC} for the quad two-input package is four times the quoted value). In some families the delay of 'AND-OR-INVERT' gates is a nanosecond or two greater than the delay of the simple 'NAND' gate.

Once a designer has selected a particular family of logic for an equipment, much of the detailed design work can be done with the aid of a simple set of figures similar to those used in Table 9.1, plus a set of simple 'engineers' design rules'. However, until a particular logic family has been selected, samples from all potential suppliers have been fully evaluated, and 'design rules' drawn up, design information should be taken from the manufacturers' full specifications.

In Table 9.1, figures in parentheses are the values at the lower and upper temperature limits. As far as possible, any significant differences in the conditions under which a parameter is measured have been noted.

Manufacturers are listed in the table only when their product differs in some essential respect from that of the first firm to introduce the family (see Section 3.2). Where no manufacturer's name is given, all available data sheets for the family quote the same figures.

9.2 Speed

9.2.1 GENERAL

The main speed parameters of all T.T.L. devices are t_{PHL}, t_{PLH}, t_{THL}, and t_{TLH}. Because the propagation delay consists of a delay period before the output starts to change, plus a portion of the rise or fall time, t_{PHL} and t_{THL} are interdependent, as are t_{PLH} and t_{TLH}. (See Figs. 5.1 and 5.2.) Propagation delays are usually measured at 1·5 V above earth on both input and output, and rise and fall times are measured either between 10 and 90 per cent of the edge or between 1 and 2 V. 10–90 per cent rise and fall time measurements may be confused by irregularities on the edges, and figures from automatic testers or digital read-out oscilloscopes may be misleading. Measurements are usually taken with a simulated load, with a standard capacity of 15pF.

9.2.2 SPECIFIED FIGURES

The S.U.H.L. 1 gate is quoted by most manufacturers as having a worst propagation delay of 20 ns (ON or OFF). Transitron do not specify a maximum for t_{PHL} or t_{PLH}, but specify a maximum propagation delay, t_{PD}, which is the average of t_{PHL} and t_{PLH} of 18 ns for all inverting gate types. Other manufacturers specify t_{PHL} and t_{PLH} separately, 20 ns for the 'NAND' gates, rising to 22 ns for the 'dual four AND-OR-INVERT' and 23 ns for the 'triple three' and 'quad two', where the extra capacity of the internal connections on the collectors and emitters of the phase-splitter transistors slows the working of the gate.

The corresponding fast gates are specified at 10, 11, and 12 ns by all manufacturers of S.U.H.L.

Expandable eight-input 'NAND' gates are made in quite different ways by different manufacturers, so there are two different figures, the worst being a t_{PHL} of 28 ns and t_{PLH} of 20 ns.

Line drivers and non-inverting gates are intermediate in speed between fast and slow gates.

Some Series 74 devices are specified with a maximum turn-on delay of 15 ns and maximum turn-off delay of 29 ns, but as manufacturers issue new data sheets, most of them are specifying a decreased turn-off delay of 22 ns. Minimum delay figures are not generally quoted for Series 54/74 devices.

Series 9000 devices have two speed specifications; 12 ns t_{PHL} and 10 ns

t_{PLH} for the military version, and 15 and 13 ns for the commercial (restricted temperature) range. Series 9000 devices also have a specified minimum delay of 3 ns t_{PHL} and t_{PLH}. Buffers are specified as being 2 or 3 ns slower than the gates on t_{PLH}, and have slightly different minimum delay figures.

Several firms quote typical speed figures, but minimum figures for delays are not generally quoted. Rise and fall times are not specified by all manufacturers. Sylvania and some others quote figures up to 8 ns measured between levels of 1 and 2 V for S.U.H.L. 1 and up to 4 ns for S.U.H.L. 2, all measured with full fan-out.

Care must be taken when reading data sheets, specifications, etc., as rise and fall time definitions may be inconsistent.

9.2.3 TYPICAL FIGURES

Measurements taken on some 2000 gates of all types from a variety of manufacturers showed that in practice the differences in speed between the various families is not quite what might be expected from the limit figures quoted in the data sheets.

All families of S.U.H.L. 1 type devices were found to work substantially faster than might be expected from their specifications. Series 54/74 did not have such a generous margin on the speed specification, so, although on paper the families might appear to offer comparable speed, in practice S.U.H.L. 1 is noticeably faster. Series 9000 devices proved to be little faster than S.U.H.L. 1, but the edges were significantly sharper (a disadvantage when cross-talk is considered).

Much less difference was noticed between S.U.H.L. 2 types and 54H/74H (only a few 74H devices were tested), and both types yielded figures much closer to the specified limit values than the 'standard' families.

For all purposes in this book, the logic families are divided into 'slow' and 'fast' depending on whether they have a diode or a second transistor in the pull-up circuit. Thus although Series 9000 is nearer in speed to S.U.H.L. 1 than it is to S.U.H.L. 2, it is nevertheless treated as 'fast' logic.

Series 54L/74L would be classed as 'very slow', but this family is not considered in any detail in this book because once a reader has learnt to handle any of the faster families, 74L will present no problems!

54S/74S and Ray 3 are included in the 'fast' category although they are substantially faster than Series 9000, 74H, or S.U.H.L. 2. In practice they may require extra care to guard against the effects of ring and cross-talk.

The speed tests carried out showed that generally all families of T.T.L. were quite consistent in speed from batch to batch, although there were the expected 'odd devices', some of which were outside the specified limits. These 'rejects' represented an insignificant percentage of the number of devices tested.

Gates in multiple-gate packages were found to have very similar speeds, and several manufacturers agreed that they would be prepared to consider a limit on the difference in delay between gates in the same package when a user's purchasing specification is drafted. It was noted that on typical complex circuits on boards up to approximately six inches square, and carrying 40–50 D.I.P.s the gates ran considerably faster (about 80 per cent of the delay for slow gates and 60 per cent for fast gates) than on the speed measurement tester with its simulated load and with the package plugged into a socket.

9.2.4 RATIO OF EDGE SPEED TO PROPAGATION DELAY

The ratios t_{THL}/t_{PHL} and t_{TLH}/t_{PLH} are not constant, but they are sufficiently close to draw some general conclusions. For slow logic, both ratios are around unity (limits found were 0·63–1·78). For fast logic the edges are considerably faster in relation to the delay, the ratios being around 0·6 for t_{THL}/t_{PHL} and 0·3 for t_{TLH}/t_{PLH}.

Slow logic fall times (10–90 per cent) were generally around 6 to 8 ns, with slightly slower rise times. Series 74 rise times were usually slower than S.U.H.L. Fast logic generally had edge speeds between 1·8 and 4 ns.

9.2.5 UNREQUIRED INPUTS ON 'NAND' GATES

The propagation delay of a gate is affected by its input connections. Each input has capacitance to the base of the input transistor VT1, and if inputs are held at a fixed potential this capacitance slows down movement of the base on switching. The fastest action is thus obtained if all inputs of an 'AND' or 'NAND' gate are strapped together and connected to the driving gate.

When a multi-input 'NAND' gate is being used and not all the input connections are required for the logical working of the gate, the unrequired inputs should be connected to the input which actually switches the gate. If the gate is turned ON by one signal and OFF by another, the connection to the unrequired input should be determined by whether t_{PHL} or t_{PLH} is the more critical. When such connections are being determined, care should be taken to observe the fan-out rules for the driving gates. If an unrequired input cannot be connected to any other input without breaking the fan-out rules then it can be left open circuit *at its pin* for the fastest working. Tests have shown that such an open circuit input will not pick up sufficient noise to switch the gate even from a fast edge from the output of fast logic.

The connection of unrequired inputs to the supply rail is sometimes advised. This should be done only if it can be guaranteed that the rail will never rise sufficiently high to risk damage to an input transistor (5·5 V). Reports have been heard of substantial quantities of logic being damaged by rail surges fed to unrequired inputs.

In any case where it is desired to set a definite '1' level, the use of a 1 kΩ resistor to the rail is advised. One such resistor can safely supply fifteen inputs at worst-case figures. In practical tests, '1' levels have been consistently set with 'open' inputs.

9.3 Fan-out

9.3.1 INTRODUCTION

The fan-out of a T.T.L. gate is determined by dividing the current handling capability of the output by the current requirements of the inputs it is to feed. Both ON state ('0' level) and OFF state ('1' level) current requirements must be considered.

The input and output currents are all defined in the specifications in relation to '0' and '1' level voltages, from which the noise margin figures are derived. An excessive fan-out will therefore result in a risk of loss of noise margin.

9.3.2 PARALLEL CONNECTIONS TO INPUTS OF ONE GATE

It is important to note that where two or more inputs to a gate are commoned, there is no increase in the '0' state input current required to be sunk by the driving gate, but in the '1' state the input leakage currents are added and the driving gate should be able to source the specified input leakage current for each input to which it is connected. If the connection of 'unrequired' inputs would increase the total fan-out beyond the limit stated for the driving gate this may be accepted *provided* that a reduction in the '1' level noise margin of about 0·01 V per excess 'unrequired' input can be tolerated. Table 9.3 shows that many of the different T.T.L. families have a considerably higher '1' level fan-out than their maximum '0' level capability.

9.3.3 SPECIFIED FIGURES

Calculations of fan-out are complicated by the existence of military and commercial specifications and the various different logic families available from different manufacturers. Table 9.2 lists the worst-case current specifications for logic gates and line drivers. If T.T.L. gates are to be used to drive loads other than normal T.T.L. gate inputs, the drive capability may be calculated from the figures in this table.

TABLE 9.2 Specified input and output currents for standard gates and buffers (line drivers)

	Currents (All values in mA)			
	Outputs		Inputs	
	0 state	1 state	0 state	1 state
54L/74L	2	0·1	0·18	0·01
54/74, Micronor 5	16	0·4	1·6	0·04
54H/74H	20	0·5	2·0	0·05
S.U.H.L. 1, Ray 1, M.T.T.L. 1 (Comm. L.F.O.)	10	0·6	1·66	0·10
H.L.T.T.L. 1 (Mil. & Comm. L.F.O.); S.U.H.L. 1, Ray 1, M.T.T.L. 1 (Mil. L.F.O.)	10	0·7	1·33	0·10
H.L.T.T.L. 2 (Mil. & Comm. L.F.O.)	11	0·75	2·2	0·15
S.U.H.L. 2, Ray 2, M.T.T.L. 2 (Mil. L.F.O.)	12	1·2	2·0	0·10
S.U.H.L. 2, Ray 2, M.T.T.L. 2 (Comm. L.F.O.)	12·5	1·0	2·5	0·10
Series 10	13	0·8	1·6	0·1
System 11	16	0·8	1·6	0·04
9000 (Comm.)	16	1·2	1·6	0·06
9000 (Mil.)	17·6	1·32	1·6	0·06
54S/74S	20	1·0	2·0	0·10
S.U.H.L. 1, Ray 1, M.T.T.L. 1 (Comm. H.F.O.)	20	1·2	1·66	0·10
S.U.H.L. & Ray Buffers (Comm. L.F.O.)	20	1·2	2·5	0·2
H.L.T.T.L. 1 (Mil. & Comm. H.F.O.); S.U.H.L. 1, Ray 1, M.T.T.L. 1 (Mil. H.F.O.)	20	1·5	1·33	0·10
S.U.H.L. & Ray Buffers (Mil. L.F.O.)	20	1·5	2·0	0·2
54/74 Buffer	48	1·2	1·6	0·04
M.T.T.L. 3	21	2·0	2·1	0·08
H.L.T.T.L. 2 (Mil. & Comm. H.F.O.)	22	1·5	2·2	0·15
S.U.H.L. 2, Ray 2, M.T.T.L. 2 (Mil. H.F.O.)	22	2·2	2·0	0·10
Ray 3 (Mil.)	22	2·2	2·0	0·05
S.U.H.L. 2, Ray 2, M.T.T.L. 2 (Comm. H.F.O.)	22·5	1·8	2·5	0·10
Ray 3 (Comm.)	22·5	1·8	2·5	0·05
H.L.T.T.L. Buffer (Mil. & Comm. L.F.O.)	32	2·4	2·5	0·2
S.U.H.L. & Ray Buffer (Comm. H.F.O.)	40	2·4	2·5	0·2
S.U.H.L. & Ray Buffer (Mil. H.F.O.)	40	3·0	2·0	0·2
H.L.T.T.L. Buffer (Mil. & Comm. H.F.O.)	48	3·6	2·5	0·2
9000 Buffer (Comm.)	48	3·6	3·2	0·12
9000 Buffer (Mil.)	48	3·96	3·2	0·12

TABLE 9.3 Fan-out table

Driven Gate	Driving Gate																								
	54L 74L	54 74	54 H 74H	S.U.H.L. 1 Comm. Low F.O.	S.U.H.L. 1 Mil. L.F.O. H.L.T.T.L. 1 Low F.O.	H.L.T.T.L. 2 Low F.O.	S.U.H.L. 2 Mil. Low F.O.	S.U.H.L. 2 Comm. Low F.O.	Series 10	System 11	9000 comm.	9000 Mil.	S45/74S	S.U.H.L. 1 Comm. H.F.O. S.U.H.L. Driver Comm. L.F.O.	S.U.H.L. 1 Mil. H. F.O. H.L.T.T.L. 1 H.F.O. S.U.H.L. Driver Mil. L.F.O.	54 & 74 Driver	M.T.T.L. 3	H.L.T.T.L. 2 H.F.O.	S.U.H.L. 2 Mil. H.F.O. RAY 3 Mil.	S.U.H.L. 2 Comm. H.F.O. RAY 3 Comm.	H.L.T.T.L. Driver Low F.O.	S.U.H.L. Driver Comm. H.F.O.	S.U.H.L. Driver Mil. H.F.O.	H.L.T.T.L Driver H.F.O. 9000 Driver Comm.	9000 Driver Mil.
54L 74L	10 11	40 88	50 110	60 55	70 55	75 61	120 66	100 69	80 72	80 88	120 88	132 97	100 110	120 110	150 110	120 266	200 116	150 122	220 122	180 125	240 177	240 222	300 222	360 266	396 266
54 74 System 11	2 1	10 10	12 12	15 6	17 6	18 6	30 7	25 7	20 8	20 10	30 10	33 11	25 12	30 12	37 12	30 30	50 13	37 13	55 13	45 14	60 20	60 25	75 25	90 30	99 30
S.U.H.L. 1 Comm.	1 1	4 9	5 12	6 6	7 6	7 6	12 7	10 7	8 7	8 9	12 9	13 10	10 12	12 12	15 12	12 28	20 12	15 13	22 13	18 13	24 19	24 24	30 24	36 28	39 28
S.U.H.L. 1 Mil. H.L.T.T.L. 1	1 1	4 12	5 15	6 7	7 7	7 8	12 9	10 9	8 9	8 12	12 12	13 13	10 15	12 15	15 15	12 36	20 15	15 16	22 16	18 16	24 24	24 30	30 30	36 36	39 36
H.L.T.T.L. 2	0 0	2 7	3 9	4 4	4 4	5 5	8 5	6 5	5 5	5 7	8 7	8 8	6 9	8 9	10 9	8 21	13 9	10 10	14 10	12 10	16 14	16 18	20 18	24 21	26 21
S.U.H.L. 2 Mil. 54S 74S	1 1	4 8	5 10	6 5	7 5	7 5	12 6	10 6	8 6	8 8	12 8	13 8	10 10	12 10	15 10	12 24	20 10	15 11	22 11	18 11	24 16	24 20	30 20	36 24	39 24

S.U.H.L. 2 Comm.	1 0	4 6	5 8	6 4	7 4	7 4	12 4	10 5	8 5	8 6	12 6	13 7	10 8	12 8	15 8	12 19	20 8	15 8	22 8	18 9	24 12	24 16	30 16	36 19	39 19
Series 10	1 1	4 10	5 12	6 6	7 6	7 6	12 7	10 7	8 8	8 10	12 10	13 11	10 12	12 12	15 12	12 30	20 13	15 13	22 13	18 14	24 20	24 25	30 25	36 30	39 30
Series 9000	1 1	6 10	8 12	10 6	11 6	12 6	20 7	16 7	13 8	13 10	20 10	22 11	16 12	20 12	25 12	20 30	33 13	25 13	36 13	30 14	40 20	40 25	50 25	60 30	66 30
S.U.H.L. Driver Comm. H.L.T.T.L. Driver	0 0	2 6	2 8	3 4	3 4	3 4	6 4	5 5	4 5	4 6	6 6	6 7	5 8	6 8	7 8	6 19	10 8	7 8	11 8	9 9	12 12	12 16	15 16	18 19	19 19
S.U.H.L. Driver Mil.	0 1	2 8	2 10	3 5	3 5	3 5	6 6	5 6	4 6	4 8	6 8	6 7	5 10	6 10	7 10	6 24	10 10	7 11	11 11	9 11	12 16	12 20	15 20	18 24	19 24
M.T.T.L. 3	1 0	5 7	6 9	7 4	8 4	9 5	15 5	12 6	10 6	10 7	15 7	16 8	12 9	15 9	18 9	15 22	25 10	18 10	27 10	22 10	30 15	30 19	37 19	45 22	49 22
54H 74H RAY 3 Mil.	2 1	8 8	10 10	12 5	12 5	15 5	24 6	20 6	16 6	16 8	24 8	26 8	20 10	24 10	30 10	24 24	40 10	30 11	44 11	36 11	48 16	48 20	60 20	72 24	79 24
RAY 3 Comm.	2 0	8 6	10 8	12 4	12 4	15 4	24 4	20 5	16 5	16 6	24 6	26 7	20 8	24 8	30 8	24 19	40 8	30 8	44 8	36 9	48 12	48 16	60 16	72 19	79 19
9000 Driver	0 0	3 5	4 6	5 3	5 3	6 3	10 3	8 3	6 4	6 5	10 5	11 5	8 6	10 6	12 6	10 15	16 6	12 6	18 6	15 7	20 10	20 12	25 12	30 15	33 15

The figures in Table 9.2 can also be used if the fan-out is made up of gates from different families. It can be seen that some buffers have input characteristics which differ from the normal gate characteristics for the family. Some flip-flops and complex elements also have higher input loadings than normal gates, and these must be borne in mind when assessing gate loadings. In all cases where there is any doubt, the manufacturer's data sheets should be consulted.

9.3.4 FAN-OUT TABLE

Table 9.3 gives the worst-case specified fan-out for every possible combination of T.T.L. gates and buffers. In each square the upper figure is the '1' level fan-out, and the lower figure is that for the '0' level. In some cases these figures are equal, but generally the fan-out achievable is limited by the '0' level values.

The lower value for each combination is printed boldly. Where this applies to the '1' level, the figure applicable to the '0' level could be used with a possible sacrifice in '1' level noise margin.

9.3.5. MEASURED VALUES

Measurements on gates from many manufacturers have shown that at the stated current levels (at $V_{CC} = 5{\cdot}0$ V, room temperature) all slow devices were well within the stated limits for V_{OL} and V_{OH}; the highest '0' level measured on a slow gate being 0·286 V, and the lowest '1' level being 3·29 V. Typical average values were about 0·18 and 3·48 V. Some fast gates were nearer the limits, and there were a few marginal rejects, but the typical average values were still well clear of the limits. Thus on typical gates, fan-out rules could be exceeded by up to about double the stated figure without risk of reducing the worst-case calculated noise margins, even if all inputs were at their worst-case current levels.

Worst input current levels measured were nearer to the specified figures for I_{IL} with some marginal rejects, but the typical average figure was around one milliampere. I_{IH} values were generally very well down on the specified figures, typically less than 10 μA. Most batches of gates contained an odd one or two inputs with leakages between about 20 and 50 μA.

The figures indicate that typically the circuits are well within their specified limits and the fan-out rules normally offer a generous safety margin, which will result in better voltage noise margins and noise immunity in use. The low values of I_{IH} measured indicate that the 'strapping' of unrequired inputs to working inputs is statistically unlikely to cause any trouble even if fan-out rules are broken.

A limited number of gates were tested to see at what current levels the output voltage would reach its specified value. It was found that S.U.H.L. 1, S.U.H.L. 2, Series 74, and Series 9000 gates would all sink over 30 mA before V_{OL} rose to 0·4 V (with V_{CC} = 5·0 V, at room temperature). The '1' level currents showed a much wider margin. The lowest current value found which pulled an output down to 3·0 V was 4 mA. It was noted that Series 9000 gates had outstandingly good '1' level current sourcing capability and the average current measured (to set V_{OH} to 3·0 V) was 16·5 mA. Because there was no automatic test equipment available for these tests, only a few packages from each family were tested, and the results should not be regarded as reliable design information.

The '1' level currents measured indicate that only in most exceptional circumstances is there any risk of noise immunity being lost because the nominal '1' level fan-out has been exceeded.

If fan-out rules are knowingly broken, it must be borne in mind that all the above figures were obtained with a 5 V supply at room temperature, and they can be expected to degrade at environmental extremes.

9.4 Noise Immunity

9.4.1 INTRODUCTION

The noise immunity of a logic gate or family of gates is a measure of the ability of the logic to operate correctly without responding to spurious signals caused by cross-talk, line reflections, or power or earth rail disturbances. Noise immunity depends on the coupling to be expected between adjacent signal lines, the input and output impedances of the logic gates, the voltage noise margins of the logic, and the duration of any spurious pulses.

The coupling to be expected between signal lines depends on the spacing between the lines and their dimensions, the edge speeds of the signals, the gate impedances, and the voltage levels on the lines. Cross-talk and reflections on T.T.L. interconnections are dealt with in Chapters 14 and 15.

Voltage noise margins can be determined from the d.c. parameters of the logic, since they are the differences between the gate output levels and the threshold levels at which the input switches. These can be determined worst case from the device specification, or, for any given device, from the low frequency transfer characteristic (Fig. 4.7). The internal delays of a saturating logic circuit can result in a gate having either a better or a worse high frequency noise margin than the low frequency or d.c. figure. A single narrow pulse of very brief duration may have a maximum amplitude well above the d.c. noise margin, but may be so short that the gate output circuit cannot respond, or alternatively, a single pulse, whose amplitude is below the d.c. noise margin, but which occurs immediately after a switching edge, may cause spurious switching because the stored base charge on a transistor in the gate has not decayed fully following the switching action.

9.4.2 SPECIFIED VOLTAGE NOISE MARGINS

9.4.2.1 '1' *Level*

The lowest specified '1' level output voltage for any T.T.L. family is 2·4 V. However, this is specified with a V_{CC} of 4·75 V (or in some cases 4·5 V), and with an input voltage to the gate of such a level that VT2 will have started to conduct—i.e., the '1' level is specified at a point on the sloping portion of the transfer characteristic. The highest specified input threshold voltage is 2·0 V, so over the entire T.T.L. range there is a worst-case noise margin of 0·4 V at the '1' level. It should be noted that in the families where V_{OH} is specified at a reduced V_{CC}, the military versions specify at a V_{CC} of 4·5 V, whereas commercial versions give the same value for V_{OH} at a V_{CC} of 4·75 V. If a 5·0 V supply is assumed, V_{OH} will be increased by 0·25 V above the lowest specified value, so a low frequency '1' level noise margin of 0·65 V could be expected.

9.4.2.2 '0' *level*

The highest V_{OL} is specified as 0·45 V (except for Series 54S/74S) and the lowest input threshold specified is 0·8 V, so the guaranteed worst-case '0' level noise margin is 0·35 V.

The above figures allow for all possible worst-case limits of temperature and rail voltage, and to achieve these figures along a chain of gates the input circuits would have to be at one limit temperature while the output circuits on the same dice would have to be at the other limit of temperature—a condition most unlikely to be met in practice!

9.4.3 PRACTICAL NOISE CONSIDERATIONS

If the above 'impossible' worst-case voltage noise margins figures are applied to a chain of T.T.L. gates it would appear that any noise signal in excess of 0·35 V which is picked up on a '0' level node should appear inverted on the succeeding '1' level node. This can occur only if the noise is picked up by all the '0' level inputs to the gate in question, since the input which is at the lowest level will hold down the base of VT1 and allow all the other inputs to cut-off.

The capacitance at the '1' level node has an effect on the propagation of noise on a preceding '0' level. Round the 'slope' of the transfer characteristic only the upper output transistor VT4 is active, so any capacity present on the output will tend to hold the '1' level signal constant, allowing VT4 to cut off if its base voltage falls. The input on the emitter of VT1 of the driven gate is already cut off, so the node capacitance can discharge only the through leakage currents, which could be as low as a couple of microamperes (or as high as a couple of milliamperes, but measurements show that the former case is the more likely limit). Thus the response of a typical '1' level output node to noise which is propagating only through VT4 can be in the order of several microseconds before the output moves down. Only when the noise on a '0' level input rises sufficiently high and for sufficiently long to start to turn on the lower output transistor VT5 is the noise really likely to be detected on the succeeding '1' level output.

The effect of noise propagating over two stages is not generally seen in practice. Another reason for this is that any positive going noise on a '1' level node has the effect of raising the node potential and cutting off the emitter follower output of VT4 on the driving gate. Since the input of the driven gate is already in a high impedance state, the interconnecting track requires very little energy to drive it up to a high voltage. Thus a typical T.T.L. circuit can present some alarming looking coupled noise signals on '1' level lines, but examination will show that this noise is positive going and cannot cause spurious switching. It can, however, have a slight effect on the speed of the circuit since a device turning ON must first discharge the output node from an abnormally high voltage.

9.4.4 MEASURED LOW FREQUENCY PARAMETERS

As was stated in Section 9.3.5, the typical T.T.L. input current levels are comfortably within their specified limits, and the output voltages also are well clear of the limits. Thus if the quoted fan-out rules are followed, an increase in '0' level noise margin of 0·25 V can be expected, and an increase of 0·4 V at the '1' level. These figures are still using the specified limits on input threshold levels. The input threshold levels are specified as 0·8 and 2·0 V, but in practice the actual switching aperture is usually just over 0·10 V wide, which leaves a further 0·55 V each side which can be added to the typical noise margins.

Thus in practice a chain of devices can have noise margins of 1·15 V at the '0' level (0·2 V V_{OL}–1·35 V $V_{IL\ max}$) and 1·25 V at the '1' level (2·8 V V_{OH}–1·55 V $V_{IH\ min}$), with the '1' level noise margin being further increased by AC effects which tend to eliminate the 'slope' at the upper end of the low frequency transfer characteristics.

9.4.5 HIGH FREQUENCY TRANSFER CHARACTERISTICS

So far, noise immunity has been discussed as if the noise were a d.c. level or long pulse (except in Section 9.4.3). In practice, most of the noise likely to cause malfunctioning will be short duration spikes, for which the transfer characteristics and noise margins are quite different.

Figure 9.1 shows the response of a typical slow gate to triangular input pulses 12 ns wide based on standing d.c. levels equal to the normal logic levels; i.e. the positive going pulses have a base line of 0·4 V and negative going pulses are from a level of about 3·5 V. For either a positive or a negative going pulse there is no significant output response until the peak amplitude of the pulse has passed through the normal d.c. threshold (transfer characteristic as shown by the broken line) by about half a volt.

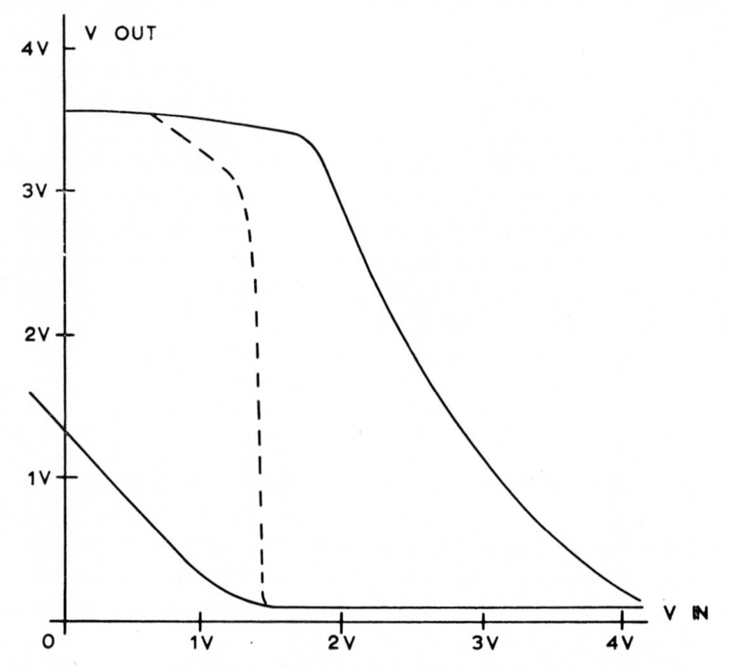

Fig. 9.1 Typical high-frequency transfer characteristic (slow logic).

A positive going input spike with a peak amplitude of about four volts will fully switch the output, but a negative going spike (from 3·5 V level) down to 0 volts will barely bring the output to the d.c. threshold level of the next gate in a chain.

The output pulse is a delayed and inverted replica of the input signal, and the high frequency characteristics are plotted at the peak points of the waveforms. Figure 9.2 shows the corresponding curves for a fast (Series 2) gate, which has about half the a.c. noise margin of the slow gate.

The amplitude of the output pulse depends on the peak amplitude of the input (as shown in Fig. 9.1) on the width of the input pulse, and on the standing d.c. level on which the pulse is based.

The effect of varying the pulse width is shown in Fig. 9.3, which shows the output voltage level achieved for an input pulse of constant amplitude whose width is measured at the normal d.c. threshold level of 1·4 V.

The first point to be noted is that the device gives much better rejection of negative going pulses than it does for positive going pulses. This is because the negative going pulse first turns the device off, then turns it on again, and, since t_{PLH} is longer than t_{PHL}, the pulse is effectively shortened,

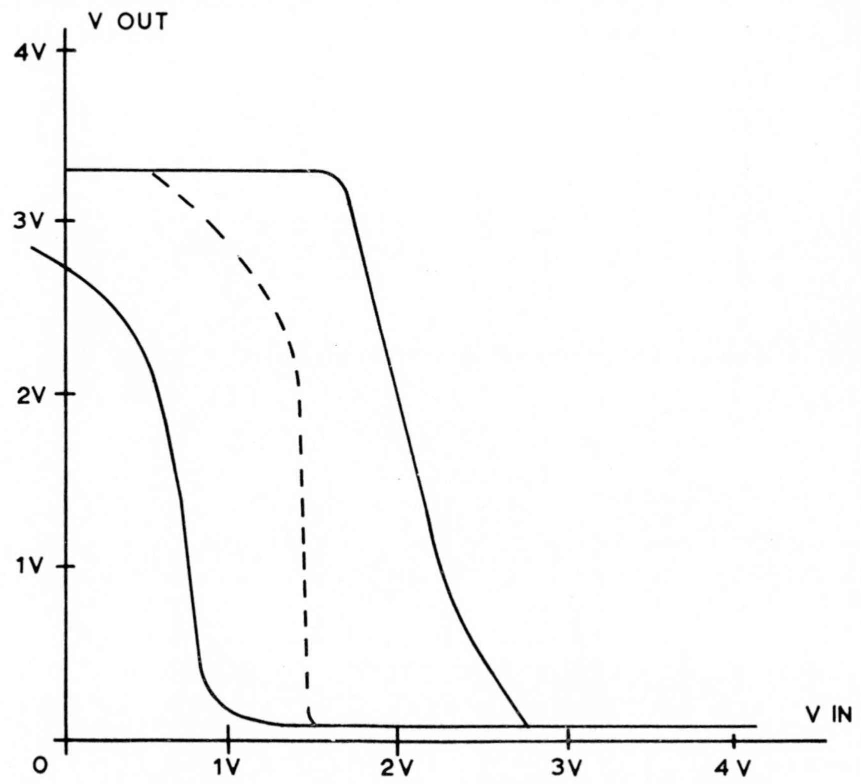

Fig. 9.2 Typical high-frequency transfer characteristic (fast logic).

whereas a positive going pulse is effectively lengthened by the difference between the turn-on and turn-off delays.

The discontinuity at about 2 V in the curve for the negative going pulses is due to the action of VT4, the 'pull-up' transistor. With short input pulses (up to about 10 ns width at the d.c. threshold level) the change in output potential is entirely due to VT5 being turned off and its collector being pulled up by the external load gate. This action ceases to be effective at higher voltages, and the width of the input pulse can be increased without much change in the resulting output level, until the input pulse is

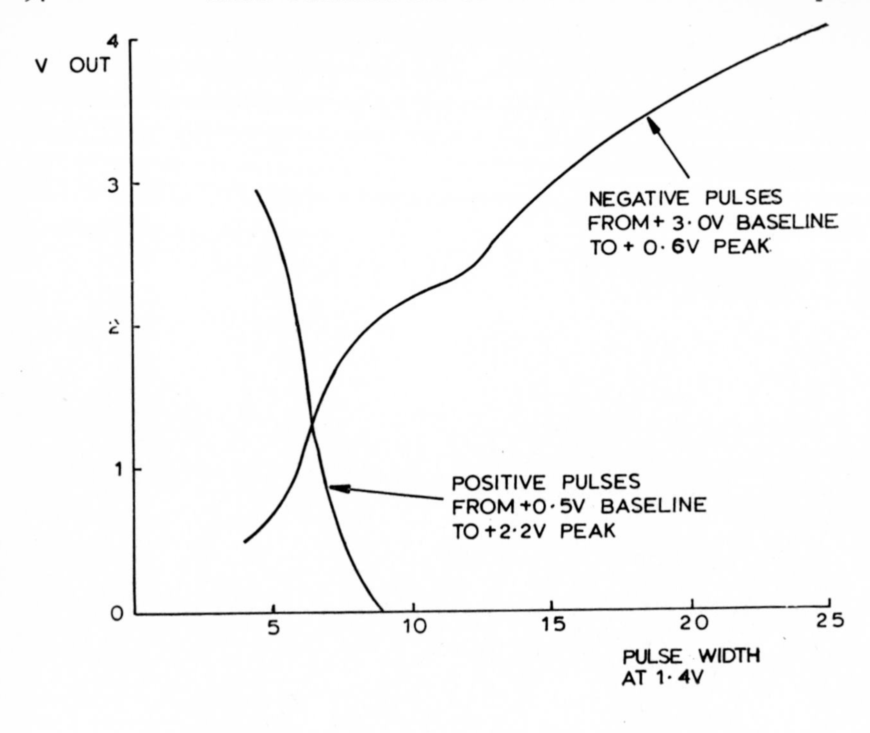

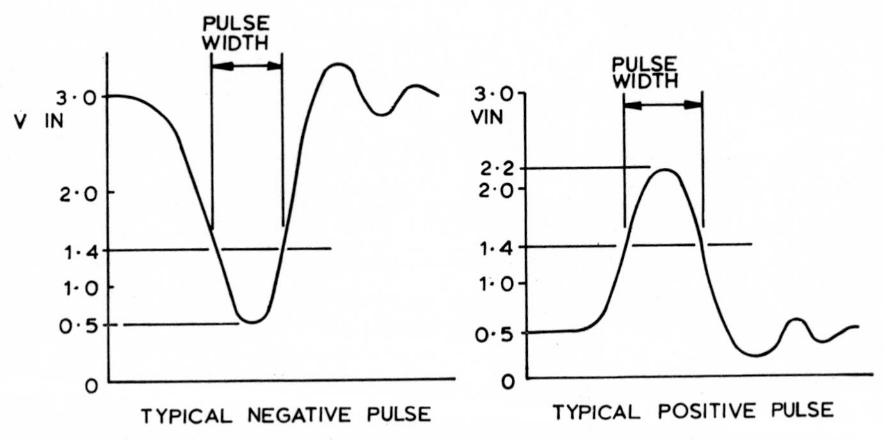

Fig. 9.3 Constant input voltage characteristics for typical slow gate. (a) Output voltage versus pulse width at threshold level. (b) Input pulse shapes.

wider than the delay through VT2, D3, and VT4, when the peak output level again rises with increasing width of the input pulse.

The effect on the peak amplitude of the output pulse of changing the d.c. baseline of the input pulse, keeping the width and peak amplitude of the input constant, is shown in Fig. 9.4, and Fig. 9.5 shows the corresponding curves for a fast gate. The effect is caused by the reduction in the pulse energy necessary to charge or discharge the base capacitances as the pulse baseline level approaches the d.c. threshold level. Here again the effect of the difference between t_{PHL} and t_{PLH} is evident in the different slopes for positive or negative going pulses.

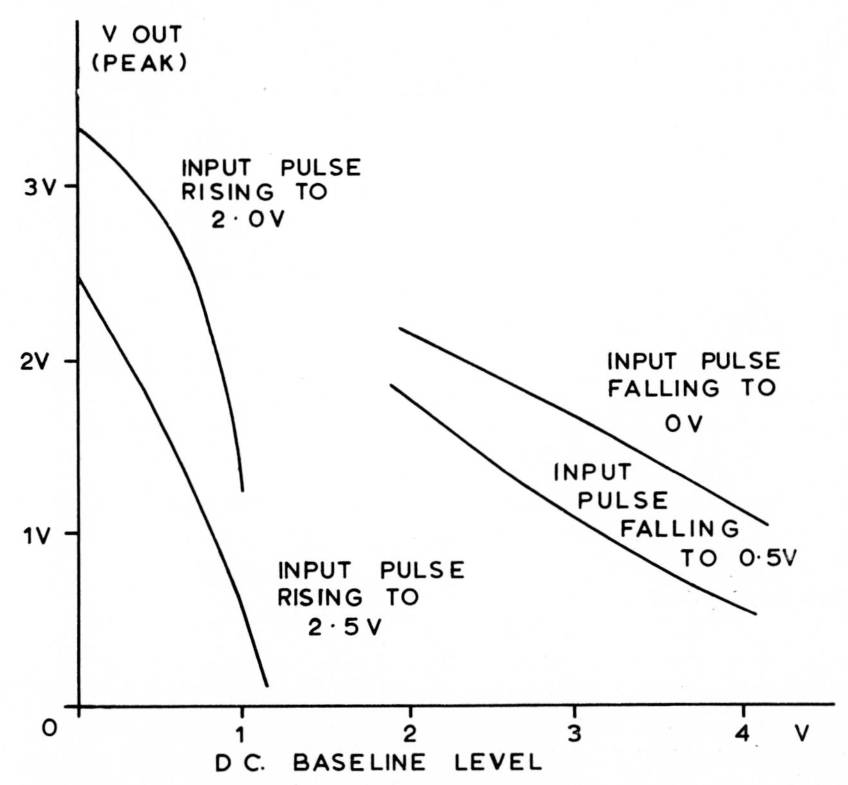

Fig. 9.4 Peak output voltage against input baseline level (12 n. sec pulse)—slow logic.

The high a.c. noise margins of T.T.L. shown by these pulse transfer characteristics mean that on most practical board layouts correct working is unlikely to be influenced by cross-talk, and if a pair of gates is cascaded with only a short interconnection the pair can be relied upon to eliminate almost completely a pulse equal in amplitude to the full logic swing and up to 12 ns in width.

Similar noise rejection performance can be expected from a gate whose input is held constant, but which is subjected to a pulse on the earth connection.

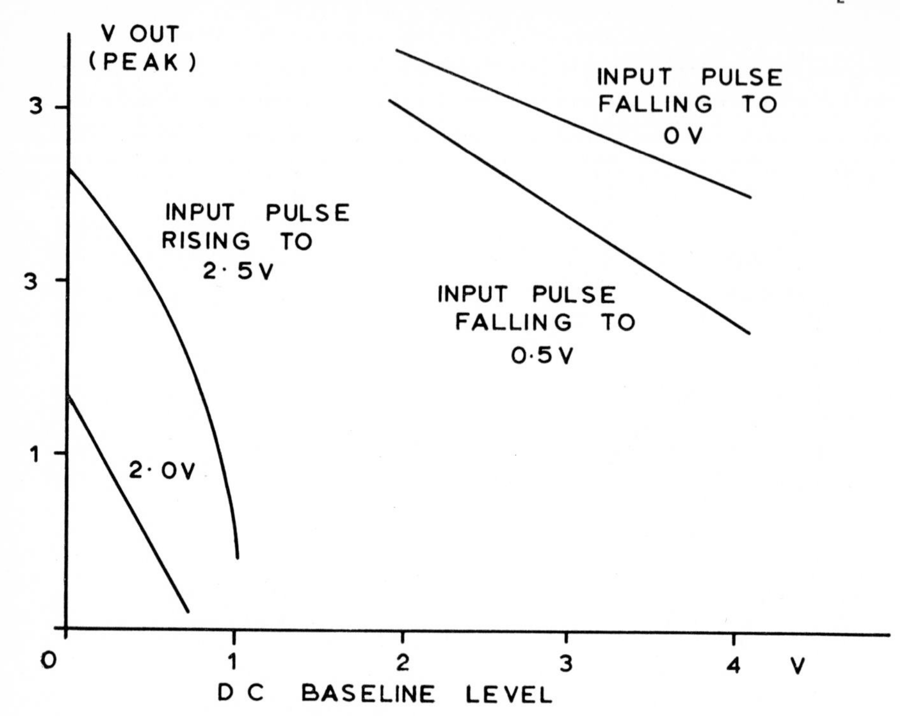

Fig. 9.5 Peak output voltage against input baseline level (12 n. sec pulse)—fast logic.

9.5 Dissipation

9.5.1 SPECIFIED AND TYPICAL VALUES

The dissipation of a T.T.L. gate varies depending on whether the gate is ON or OFF. The ON dissipation, with the inputs to a normal inverting gate open circuited, is higher than the OFF. For simple gates, the package dissipation depends on the number of gates in the package, a 'quad' package having twice the dissipation of a 'dual' under similar conditions.

Table 9.1 lists the specified '0' and '1' level gate currents. It should be noted that in some cases these are typical figures and not limits. Measurements have shown that the specified limits have a generous margin, currents in the OFF state averaging about half the specified value, and in the ON state about two thirds. Typically the low frequency a.c. dissipation was found to be 15–16 milliwatts *per gate* for S.U.H.L. 1 and 9000 'NAND' gates and 9–11 mW for Series 74. S.U.H.L. 2 has about 50 per cent higher dissipation than S.U.H.L.1.

As explained in Section 5.3, the dissipation can increase with the frequency of switching of the device. On some circuits built it has been found that the total current consumption actually falls very slightly when the circuits are running. This is because more devices are ON than are OFF in the static state.

9.5.2 BOARD OR UNIT DISSIPATION

Overall power consumption for a board or unit can be determined from the I_{CC} figures for the packages, but individual device junction temperature calculations must also take into account the load current sunk by the device outputs. Worst case, a quadruple gate package could be sinking a total of 64 mA at 0·45 V—i.e. 29 mW, which is not an insignificant percentage of the worst-case dissipation due to I_{CC}.

When individual device dissipations are being considered, it must be borne in mind that only one ON gate can sink the full input current from any multi-input T.T.L. NAND gate. Other ON gates connected to the same device may tend to share the current or they may not sink any current at all. All other things being equal, the gate driving into the smallest fan-out will be the most likely to sink all the current from a multi-input gate.

Because of the uncertainty in 'current sharing' and logic '0' levels, board or unit dissipations should be calculated by summing all T.T.L. device I_{CC}s, and then taking any resistors or other discrete loads as being connected between the supply rail and earth (with due allowances for duty cycles, etc.).

9.5.3 CURRENT SURGES

When a board or unit is first switched on, the current taken can be higher than the normally calculated value. All the output node capacitances have to be charged to their working levels, and although these charging currents may appear insignificant when they are considered individually, they can be appreciable when they are all added together for a complete unit or system.

It is also possible for excessive currents to flow during the initial switch-on of an equipment if VT4 and VT5 can be ON together. This condition can occur when the supply rail voltage is such that VT5 and VT2 can both be ON, but the current through R2 is insufficient to saturate VT2. This condition is largely a function of the values of R2 and R3 and the h_{FE} of VT2. Figure 9.6 shows a T.T.L. gate with a V_{CC} of 2·5 V. Semiconductor drops of 0·7 V are assumed, and values of resistors are taken at 4 kΩ for R1, 1 kΩ for R2 and R3, and 150 Ω for R4.

In the left-hand diagram h_{FE}VT2 is assumed to be 6. VT5 is on the point of conduction, and a very slight increase increase in V_{CC} will provide base drive to VT5.

When the voltage at the lower end of D3 is considered, it is found to be 2·0 V below V_{CC}; i.e. 0·5 V above the emitter of VT5. Thus, as soon as V_{CC} rises to provide base drive to VT5, current can flow through R4, VT4, D3, and VT5 if $V_{CE(sat)}$ VT5 is less than 0·5 V. The base of VT4 will be brought down to 1·4 V above V_{CE}VT5, so VT4 will be able to draw a fairly heavy base current through R2, so as to increase the voltage drop across R2 until IR2 + V_{BE}VT4 + VD3 + V_{CE}VT5 = 2·5. VT4 can thus pass a fairly high current into D3 and VT5. In the case illustrated R4 will limit this current to about 4·7 mA (depending on the $V_{CE(sat)}$ of VT2 and VT5).

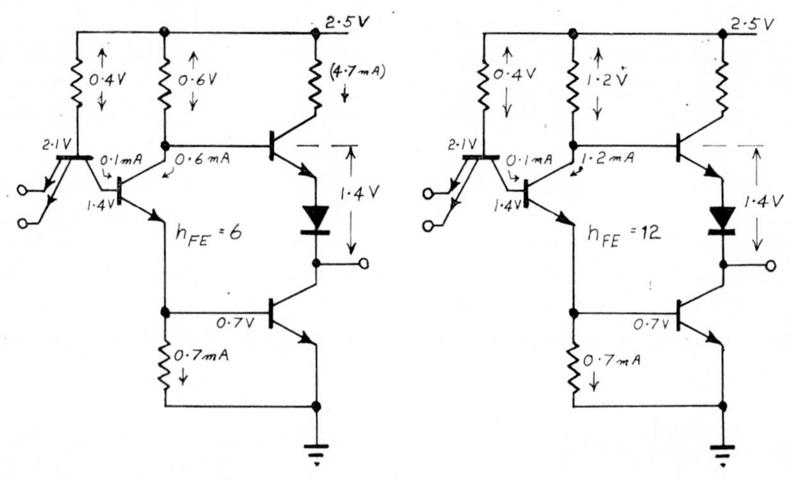

Fig. 9.6 Current conditions in T.T.L. gate while supply voltage is rising.

This unwanted conduction through VT4 can be eliminated by raising the value of R2. To guarantee that VT4 cannot conduct, about 1·0 V must be dropped across R2, and this, with a current of 0·6 mA, requires the value of R2 to be raised to 1·7 kΩ.

Alternatively, h_{FE}VT2 can be raised as shown in the right-hand diagram. This increase in h_{FE} from 6 to 12, enables VT2 to pass 1·2 mA, which brings the voltage across R2 plus that across VT4 and D3 to a figure at which VT4 cannot pass any current into VT5.

The would-be user of T.T.L. can do nothing to eliminate this surge current on a rising V_{CC} if it occurs on the logic he wishes to use, but before committing himself to a design and a logic family he should check whether this effect is present on the logic he is contemplating using.

9.5.4 UNUSED GATES

When multiple-gate packages are used ('quads', 'triples', or 'duals'), it is possible that not all the gates in the package will be used. Unit dissipation will be minimized if one input to every such unused gate is earthed. Similarly, when dual flip-flops or M.S.I. element packages are only half used, the unused half should be 'held' in its lowest dissipation state.

Such 'locking' of unused devices minimizes dissipation, and it also reduces the (very slight) risk of 'on the chip' coupling through elements left 'floating'. If board layout personnel are given a firm instruction to earth one input of every unrequired 'AND' or 'NAND' gate, there is no risk of the output of an 'AND-OR-INVERT' gate being inhibited because one of the inputs to the 'OR' function is not used and has inadvertantly been left at a '1' level. (See Section 8.2.2.)

10

Environmental Variations in Parameters

This section deals with the effects on the parameters described in Chapter 9, and on the current spike, of changes in temperature, interconnection capacitance, supply voltage, fan-out, and the use of expanders. The numbering of this section follows that of Chapter 9, so that Sections 10.?.2 deal with effects on speed. Section 10.?.1 will provide a general introduction where necessary. Table 10.1 shows the main effects on all the parameters, ignoring any second-order effects.

10.1 Temperature

10.1.1 INTRODUCTION

All the main parameters were described at a temperature of 25°C. Unless otherwise stated this section will describe the variations of main parameters between the commercial temperature limits of 0°C and 75°C.

10.1.2 EFFECT OF TEMPERATURE ON SPEED

An increase in chip temperature increases the storage times of all transistors, increases the gain, and gives greater saturation. Resistor values are increased, giving very slightly lower current levels to be switched, so the overall effect is to sharpen the edge speeds and increase the delay times of the transistors. The increase in delay times has very little effect on t_{PHL}, which is thus decreased with a rise in temperature, whereas t_{PLH} which is more dependent on storage delays is increased. These opposite effects are substantially equal in magnitude, and the overall effect on the speed of a chain of cascaded gates can be ignored. Typically the increase in t_{PLH} and the decrease in t_{PHL} can be expected to be 0·04 ns per degree centigrade rise. Figure 10.1 shows graphs of the variations of t_{PHL} and t_{PLH} with temperature. These graphs show the limits of spread of all information supplied by the device manufacturers. Experimental measurements showed that occasionally a gate may be found on which the speed varies by slightly less than the predicted amount, but none were found with more variation than shown by the worst slope drawn on the graphs.

TABLE 10.1 Simplified summary of main effects of environmental variations on major parameters of T.T.L. 'No effect' means no significant effect.

Effect of:	Temperature	Capacitance	Rail variation	Fan-out	Expanders
Effect on Speed					
t_{PHL}	−0·04 ns/°C	+0·07 ns/pF	−4%/0·5 V	Can be ignored	−2 ns
t_{PLH}	+0·04 ns/°C	+0·07 ns/pF	−4%/0·5 V		−2 ns
Fan-out	No effect	No effect	Can be ignored		No effect
Noise immunity					
d.c. '0' level	Decrease	No effect	No effect	See Section 9.3	No effect
d.c. '1' level	Increase	No effect	Increase		Decrease
a.c.	Decrease	No effect	No effect		Decrease
Dissipation	Can be ignored	Increase (slight)	Obvious	See Section 9.5	Minor increase
Current spike	Increase (slight)	Increase	Increase (slight)	Decrease	No effect

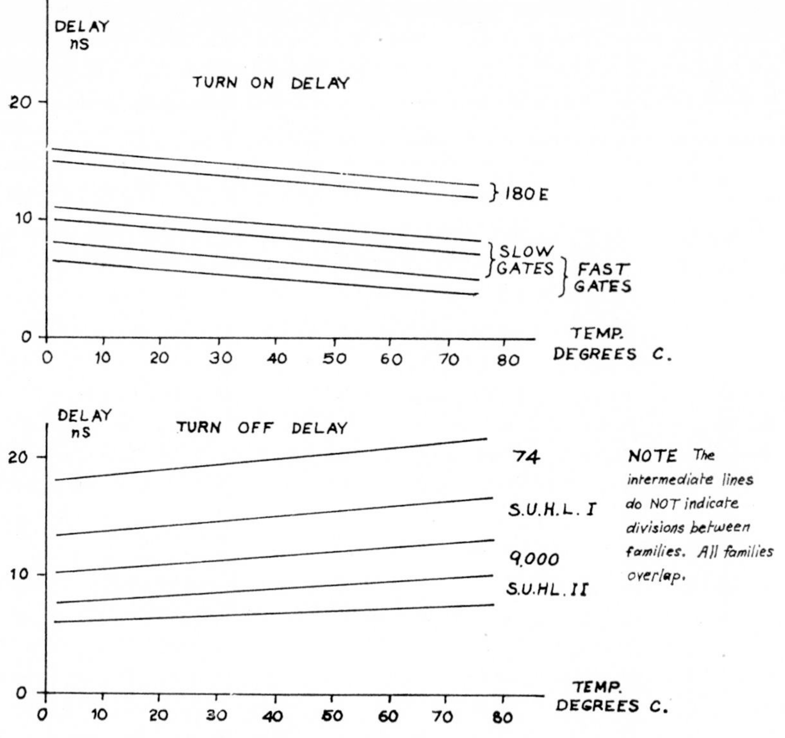

Fig. 10.1 Variation of propagation delay with temperature.

10.1.3 EFFECT OF TEMPERATURE ON FAN-OUT

Since temperature affects transistor gains and resistor values, some effect on fan-out might be anticipated, but the specification figures of currents from which fan-out is derived apply to all temperatures within the range, so any changes in actual current values can be ignored.

10.1.4 EFFECT OF TEMPERATURE ON NOISE IMMUNITY

The d.c. noise margins at any inter-gate connection are determined by the output voltage levels of the driving gate and the input threshold voltage of the driven gate.

V_{OL} is determined by the saturation voltage of the main output transistor. As the temperature rises there will be a very slight decrease in the current which this transistor has to sink, and there will also be a slight change in $V_{CE(sat)}$. These changes can safely be ignored.

V_{OH} is determined mainly by the supply rail voltage, the drop across the diode D3, and the base to emitter drop across the pull-up transistor VT4. The voltage across R2 also determines the output voltage, but since VT2 is off when the output is '1' the current flow through R2 will be small, and the voltage drop across R2 can be ignored unless the gate has expanders connected which increase the leakage currents.

A rise in temperature will decrease the semiconductor drops by about 0·002 V per degree centigrade, so V_{OH} will tend to rise with increasing temperature. As the temperature rises, leakage through VT2 (and any expanders) will increase, increasing the voltage drop across R2. This increased voltage drop across R2 will tend to decrease V_{OH}, but in an unexpanded gate the drop will usually not be sufficient to offset the rise in V_{OH} caused by the decrease in the two V_{BE}s. The overall effect is that V_{OH} rises with increasing temperature, at a maximum rate of rise of 0·004 V per degree, the rate of rise falling off as the temperature rises.

The input switching threshold voltage is determined by $V_{CE(sat)}$ of VT1, and V_{BE} of VT2 and VT5. The major effect of a rise in temperature will be to decrease the V_{BE}s, so the input switching threshold will fall by about 0·004 V per degree centigrade.

Thus with rising temperature the low frequency '0' level noise immunity will decrease and the '1' level noise immunity will increase.

The effect of a rise in temperature on the high frequency noise immunity will be determined by the effect of the change in temperature on the speed of the device (see Sections 9.4.5 and 10.1.2) and will be to decrease the immunity to positive going spikes from a '0' level and to increase the immunity to negative going spikes from a '1' level.

10.1.5 EFFECT OF TEMPERATURE ON DISSIPATION

Changes in temperature will alter slightly the current levels in the chip, but for practical purposes the effects of a change in temperature on device dissipation can be ignored.

10.1.6 EFFECT OF TEMPERATURE ON THE SWITCHING SPIKE

Changes in temperature will change the storage delay of the transistors, and so an increase in temperature will cause VT5 to pass current for a longer period during turn-off, which will increase both the amplitude and duration of the current spike. The increase in amplitude is the result of the later turning-off of VT5. This increase in the current spike energy with rising temperature is small and can generally be ignored.

10.2 Capacitance

10.2.1 INTRODUCTION

This section considers only the effects of capacitances likely to be found in normal interconnections or when driving discrete component circuits with small capacities on their inputs; i.e. up to about 100 pF. If T.T.L. gates are to be used driving into much larger capacitances, series resistors should be incorporated to limit the discharge current on turn-on to a maximum of about 20 mA.

10.2.2 EFFECT OF CAPACITANCE ON SPEED

Capacitance on the output of a T.T.L. element increases the delay of the element (measured at threshold level) and slows down the pulse edges. The rising or turn-off edge is slowed by about 0·03 ns/pF, and the falling or turn-on edge by about 0·01 ns/pF. The effect of capacity on edge speed increases at higher values of capacity.

Both turn-on delay and turn-off delay are increased by about 0·07 ns/pF. Figure 10.2 shows graphs of turn-on and turn-off delay against capacitive loading. These graphs are limit cases derived from all manufacturers' published information on *typical* figures. All devices tested were found to give results either within the lines or (in a few cases only) faster at 100 pF than predicted by the graphs. A rejected gate which was too slow to meet the speed specification showed a similar slope when tested at 15 and 100 pF loading.

10.2.3 EFFECT OF CAPACITANCE ON FAN-OUT

Fan-out is not affected by capacitance on the interconnections, but leakage currents through any capacitors may have to be considered, and if there is any such leakage, this current should be deducted from the output source current quoted in Table 9.1 and the fan-out should be re-calculated.

10.2.4 EFFECT OF CAPACITANCE ON NOISE IMMUNITY

The noise margins of T.T.L. are not affected by capacitance on the interconnections, but the coupling between adjacent tracks could be affected. (See Chapter 14.)

10.2.5 EFFECT OF CAPACITANCE ON DISSIPATION

Since capacitance on the output of a device increases the duration of the current spike on switching (see Section 10.2.6) the effect of capacitance

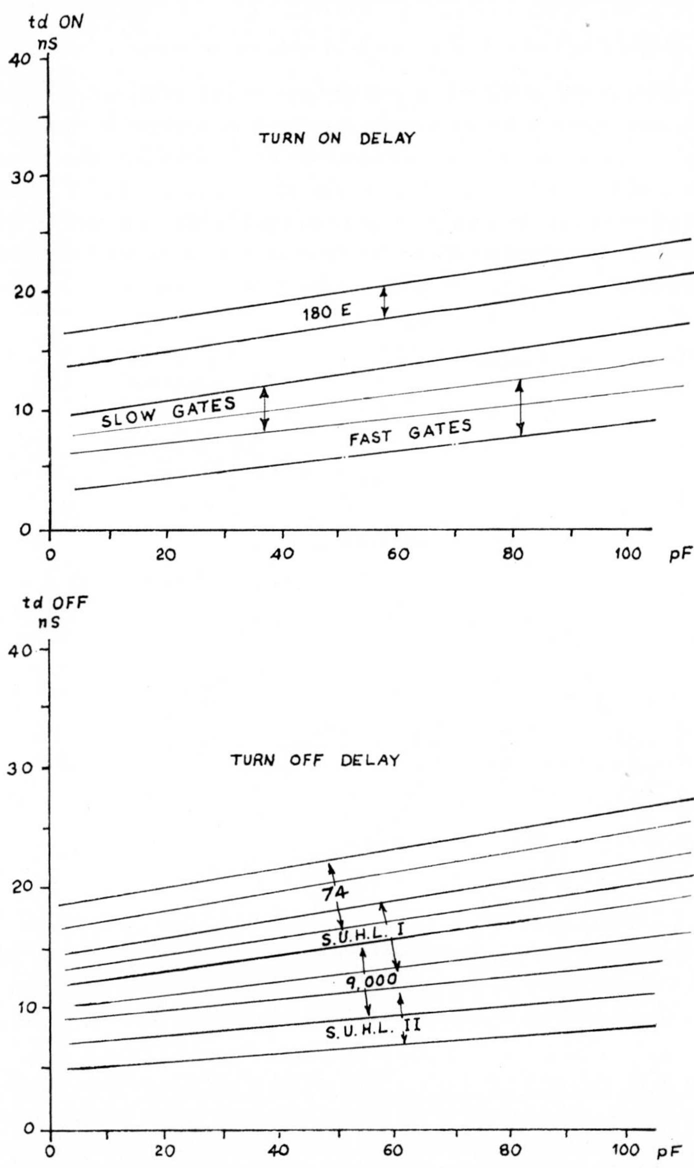

Fig. 10.2 Variation of propagation delay with capacitive loading.

will be to increase the device dissipation, the amount of the increase depending on the frequency at which the gate is switched, the device characteristics, and the value of the capacitive load (see Section 5.3). For most practical purposes this effect can be ignored (i.e. below 5 MHz switching frequency).

10.2.6 EFFECT OF CAPACITANCE ON THE SWITCHING SPIKE

The effect of the addition of capacitance to the collector OR-expansion point in generating a turn-on spike has been described in Section 6.1.3. Any such capacitance on the collector of VT2 will delay the rise of the lower end of R2 when the gate is turned off. The turn-off of VT5 will not be delayed so the turn-off current spike will occur later and will be reduced in amplitude. Capacitance on the emitter of VT2 will tend to hold VT5 on during turn-off and will thus increase the duration and hence the

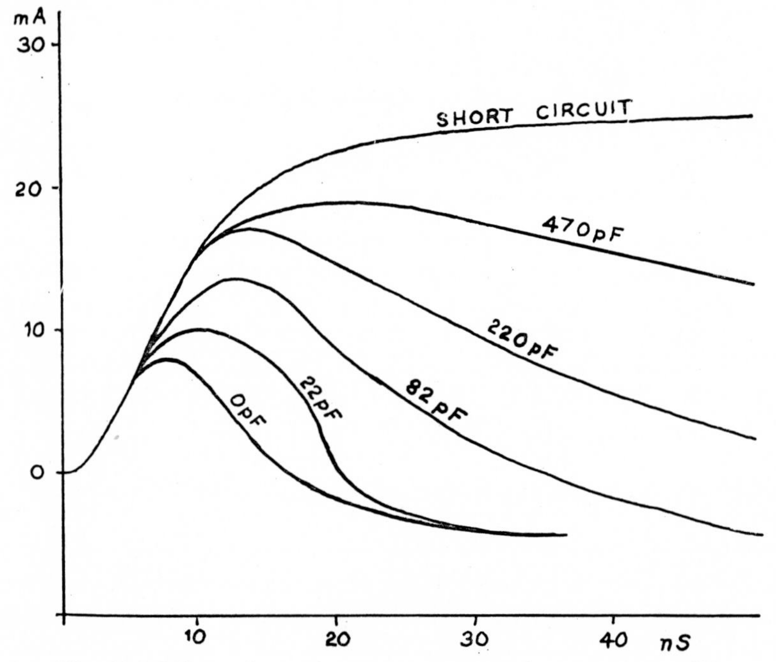

Fig. 10.3 Effect of capacitance at output node on current spike.

amplitude of the current spike. This effect is not so great as the effect of capacity on the collector of VT2, and the average effect of the addition of an expander is to reduce the energy content of the turn-off spike.

Capacitance on the output node has to be charged by the current through VT4 before the potential of the output can rise, and this charging current starts to flow as VT5 desaturates and the output potential begins to rise. For small values of capacitance the effect is to extend the duration of the current spike up to about 20 ns, then as the capacitance rises above 30–40 pF the amplitude of the current spike rises and the peak current occurs later, until at about 500 pF the maximum current is approximately equal to the output short-circuit current of the device (see Fig. 10.3) and takes over 100 ns to decay to the normal OFF level.

When a full short circuit is applied to the output, the current rises to about two thirds of its final value in 10 ns, and rises to I_{CC} OFF plus I_{OS} in about 40–50 ns.

10.3 Supply Rail Variation

10.3.1 INTRODUCTION

The effects of rail variation are considered between 4·5 and 5·5 V. T.T.L. devices will usually continue working between much wider limits (3·1–7·5 V in one known case) but voltages outside the 5·0 V plus or minus 0·5 V should not be considered for use in fully toleranced designs. Several manufacturers specify their devices only between 4·75 and 5·25 V, especially commercial varieties.

10.3.2 EFFECT OF RAIL VARIATION ON SPEED

An increase in rail voltage increases base drives available to the transistors, and gives faster switching. Generally the increase in speed is more noticeable on the turn-off delay than on the turn-on delay. Individual device tests indicated typical variations as shown in Fig. 10.4, but 'on board' tests of circuits with many gates switching in sequence indicated that an overall figure of −8 per cent per volt should be applied to the total t_{pd}s over a complete system or chain of gates.

10.3.3 EFFECT OF RAIL VARIATION ON FAN-OUT

10.3.3.1 *Output of the driving gate*

(a) '0' level. The output transistor VT5 is normally heavily saturated at its full rated load (see Section 3.2) and the loss of base drive due to a 10 per cent decrease in rail voltage will not significantly affect the ability of the output transistor to sink the rated current while maintaining V_{OL} below its specified level.

(b) '1' level. A decrease in rail voltage will give a decrease in V_{OH} such that a device which is on or near the specified limits for its '1' level output could not source the rated current with a supply voltage of 4·5 V. Therefore the effect of changes in rail voltage on the '1' level fan-out must be ignored, and the entire effect of the reduced rail voltage must be considered as affecting the '1' level noise immunity. (See Section 10.3.4.)

10.3.3.2 *Input of the driving gate*

Only the '0' level need be considered (see above), and this is very nearly a one to one effect—i.e. a change in rail voltage of 10 per cent will change

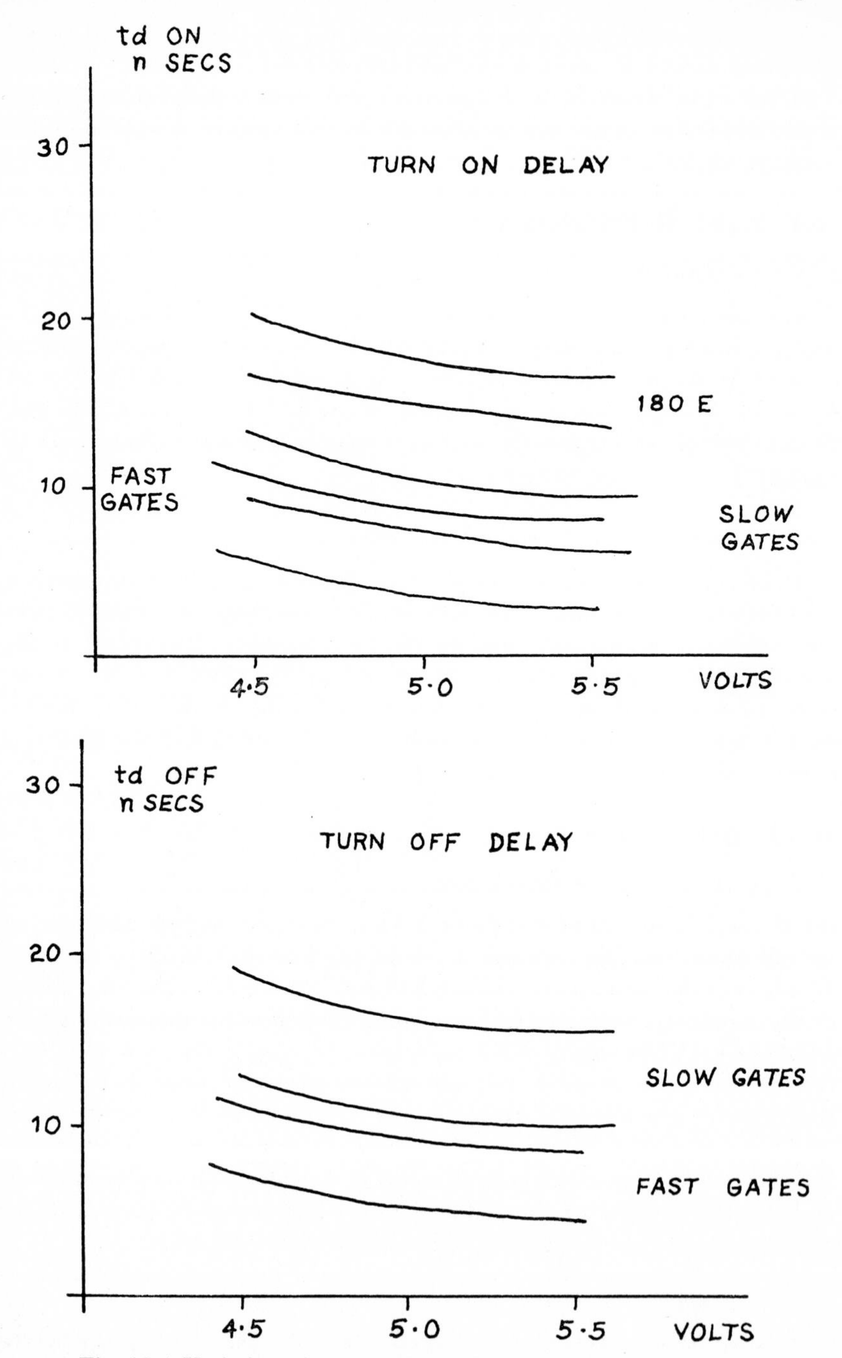

Fig. 10.4 Variation of propagation delay with supply rail voltage.

the input current by about $11\frac{1}{2}$ per cent. If the rail voltage increased, this increase in input current could require a decrease in the fan-out allowed. However, in some specifications the input current is measured and specified at 0 V, whereas the driving gate output is specified at 0·45 V. This means that in the basic fan-out calculations in Section 9.3 the '0' level fan-outs quoted are pessimistic by about 10 per cent.

In practice the difference between the $11\frac{1}{2}$ per cent increase in input current when the supply voltage is 0·5 V high and the decrease in input current of 10 per cent when the input is taken to 0·45 V instead of to 0 V can be ignored, and the fan-outs quoted in Section 9.3 can be applied over the working range of supply voltages from 4·5 to 5·5 V. In the specifications where the input current is measured at V_{OL}, the highest toleranced rail voltage is applied for the measurement, so again the figures specified in Section 9.3 are pessimistic and can be applied over the full temperature range.

10.3.4 EFFECT OF RAIL VARIATION ON NOISE IMMUNITY

As described in Section 10.3.3, variation in the supply rail voltage has no significant effect on a '0' level output. Because the input switching threshold voltage is determined by voltage drops across semiconductors (see Section 10.1.4) this voltage will not be affected significantly by variations in V_{CC}. However, the '1' level output is determined largely by semiconductor voltage drops from V_{CC}, so any (low frequency) variation in the supply rail voltage will appear as an equal variation in the '1' level output voltage (but see Section 9.3.4).

Thus if the rail voltage (to the gate driving the node considered) is decreased by 0·5, then the '1' level noise margin at the driven inputs will also be decreased by 0·5 V. '0' level noise margins are not affected by changes in rail voltage, and there is no significant direct effect on the high frequency noise immunity. However, since the change in rail voltage affects the speed of the device, there will be an indirect effect as the faster the operation of the device the less will be the high frequency noise immunity. An increase in rail voltage will thus give a small decrease in high frequency noise immunity, which, at the '1' level will be more than compensated for by the direct increase in the low frequency noise margin.

10.3.5 EFFECT OF RAIL VARIATION ON DISSIPATION

The current taken from the supply rail varies almost directly as the supply voltage varies—i.e. the gate behaves very nearly as a resistor, so dissipation will increase with a rising rail voltage, following the usual square law.

10.3.6 EFFECT OF RAIL VARIATION ON THE SWITCHING SPIKE

An increase in the supply rail voltage will increase the current levels within the gate. The current through R3 is set by V_{BE}VT5, so the increased current through VT2 is available as increased base overdrive to VT5. This increases the stored base charge which has to be swept out by the current spike, so the effect of an increase in rail voltage will be a slight increase in the amplitude and duration of the current spike.

10.4 Fan-out

10.4.1 INTRODUCTION

The effects on any parameters caused by changing the fan-out being driven are very difficult to determine as a change in fan-out usually involves changing the capacitance and inductance of the interconnections, and the effect due to the change in capacitance will usually hide any effect due to the change in fan-out.

10.4.2 EFFECT OF FAN-OUT ON SPEED

The effect of fan-out on speed can be ignored when designing T.T.L. circuits. If changes in fan-out could be made without affecting the capacitance, the effect would be a slight increase in t_{PHL} with increasing fan-out and a decrease in t_{PLH}, the overall result of which would be a very slight decrease in t_{pd}.

10.4.3 EFFECT OF FAN-OUT ON FAN-OUT

(Cannot exist.)

10.4.4 EFFECT OF FAN-OUT ON NOISE IMMUNITY

See Sections 9.3 and 9.4 for the effect of fan-out on noise margins. The major effect of a large fan-out on noise immunity (at the inputs of the devices being fed by the heavily loaded gates) lies in the increase in track length likely to be necessary to interconnect the large fan-out, which might result in increased coupling and cross-talk (see Chapter 16).

10.4.5 EFFECT OF FAN-OUT ON DISSIPATION

The dissipation of an ON gate may increase with an increase of fan-out, but the overall board or unit dissipation will be the sum of the individual gate I_{CC}s multiplied by the rail voltage regardless of the individual fan-outs used (see Section 9.5).

10.4.6 EFFECT OF FAN-OUT ON THE SWITCHING SPIKE

A gate driving a high fan-out will have a high current flowing through VT5 from the external load. This current will add to the current through VT4, which causes the current spike, and will desaturate VT5 slightly earlier, thus reducing both the peak amplitude and the duration of the supply rail current spike. Thus the spike will be worst for devices driving low fan-outs. However, the effect of the extra capacitance involved in the connections to a high fan-out can often be greater than the direct effects of the increased fan-out.

10.5 Expanders

10.5.1 INTRODUCTION

Most of the adverse effects on the performance of an expandable gate which has an expander package connected are caused by, or are amplified by, the capacitance of the expander package and the tracks which interconnect the expander points. Since these capacitances are an unavoidable feature of the use of expanders the effects of capacitance at expander points are considered in this section instead of in Section 10.1. Whenever expanders are used the tracks interconnecting the expander points must be kept as short as possible.

Since the capacitance on a normal expander-point short interconnection would never be large, it can be assumed that the capacitance is of such a value that it can be discharged virtually instantaneously by a transistor turning ON—i.e. the discharge current of the capacitor will not become a significant percentage of the current which the transistor is capable of switching.

10.5.2 EFFECTS OF EXPANDERS ON SPEED

10.5.2.1 'OR' *Expanders on inverting gates*

The main effect on speed of adding an 'OR' expander to an inverting gate is indistinguishable from the effect of adding capacitance at the expander points. The addition of capacitance at the emitter expander point (i.e. the emitter of VT2) slows down the turn-on of the gate as the current turned on by VT2 has to charge the capacitance before it can provide any base drive to VT5. The turn-off delay is not affected by capacity on the emitter expander point because the potential across the capacity has to drop by only a millivolt or two to stop base drive to VT5. The addition of capacitance on the collector expander point (i.e. the collector of VT2) has no effect on the turn-on delay as only the emitter circuit of VT2 is involved in turn-on, but turn-off delay is increased. The increase in turn-off delay with capacitance on the collector is very much greater than the increase in turn-on delay with the same capacitance on the emitter.

The effects which occur when the capacitance is connected across both expander points instead of between one expander point and earth are similar. There is very little change in the degradation of turn-on delay because VT2 collector does not start to come down in voltage until after the emitter has risen, so the capacity on the emitter is still connected to a (substantially) fixed voltage point; but the effect on turn-off delay is greater when the capacitor is connected across the expander points than when it is connected from collector to earth because the emitter of VT2 is falling in voltage while the collector is trying to rise.

Typically, the addition of an expander to a gate adds about 2 ns to the average propagation delay, with the major effect being on t_{PLH}. The addition of two expanders in the same package, connected in parallel, does not double this effect since the actual interconnections remain the same. An increase of about 3 to 3·5 ns should be allowed for any such double expansion. If two separate expander packages are used the addition to the delays may be more than double that for a single expansion and 5 ns should be allowed. The figure achieved depends on the wiring capacitance, so exact figures cannot be quoted. The same increases in delay apply whether the delay is measured from an input of the expanded gate to the output or measured through the expander package. The expander itself has very little inherent delay, and speeds through the expander package to the output are usually very nearly the same as those from an input on the expanded gate. Figure 10.5 shows typical degradations in gate delays with capacitance at the expander points.

10.5.2.2 'OR' *Expanders on non-inverting gates*

The non-inverting expander has only the single expansion point on the collector of VT8 (see Fig. 7.1). The addition of capacitance at this point has only a small effect on the turn-on time of VT8, but slows down its turn-off. The output stage (VT2 to VT5) inverts the signal at the collector of VT8, so the effect of adding an expander to a non-inverting gate is a slight increase in turn-off delay (i.e. output going high), and a larger increase in turn-on delay. Typically, average propagation delay will be increased by 0·5 ns for each expander added. Edge speeds are not significantly affected.

10.5.2.3 'AND' *Expanders on eight-input gate*

Here again the major effect on the speed of the gate is caused by the capacitance of the expander and its connections to the gate package. The capacitance is added to the base and collector of VT1. When an input rises from 'low' to 'high', both these capacitances have to be charged

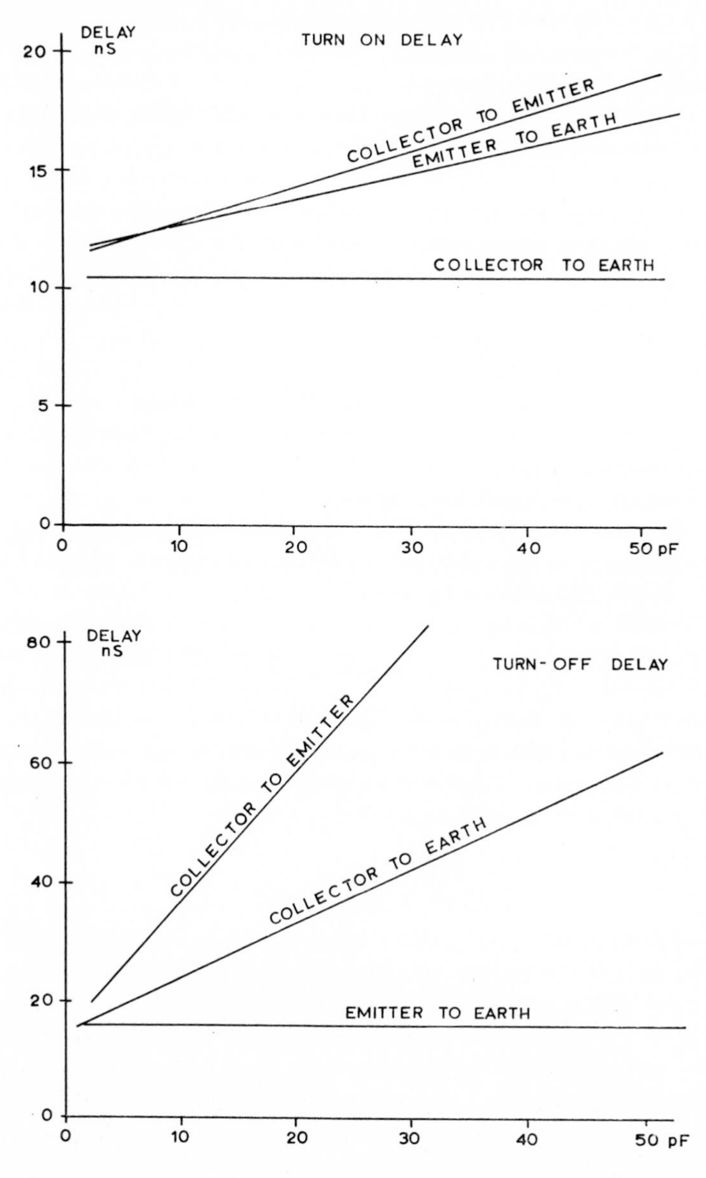

Fig. 10.5 Variation of propagation delay of inverting gate with capacitance at 'OR' expansion points.

through 4 kΩ resistor R1, so the turn-on delay will be increased by about 4 ns/pF (total capacitance added—i.e. if equal capacitance is added to base and emitter, the figure is 8 ns/pF on each capacitor). Turn-off delay is increased, but by a smaller amount, since the capacitance on the base is discharged through the emitter of VT1 and the ON output transistor of the driving gate.

When an expander is connected to the gate, some of the wiring and chip capacitance on each interconnection will be to earth, but some of the capacitance is to the other interconnection, and this latter capacitance has very much less effect on the speed of the gate than the capacitance to earth because the voltage changes on the base and collector of VT1 are both in the same direction.

Typically, the addition of one expander with packages adjacent and expander point interconnections less than one inch long will increase the average propagation delay by about 3 ns. The major effect is on t_{PHL}. Edge speeds are not significantly affected.

As in the case of the inverting 'OR' expander, the delay through the expander inputs is substantially the same as that through the main gate inputs when the expander is connected.

10.5.3 EFFECTS OF EXPANDERS ON FAN-OUT

The inverting 'OR' expander can affect the '1' level output of the expanded gate, but this effect can be considered as affecting the noise immunity at the normal fan-out. Apart from this, the addition of expanders has no effect on fan-out.

10.5.4 EFFECTS OF EXPANDERS ON NOISE IMMUNITY

This section considers the effects of the addition of expanders on the noise immunity at the normal gate output nodes. The noise immunity of the expander points themselves is considered in Section 7.4.

10.5.4.1 'OR' *Expanders on inverting gates*

The addition of expanders to an inverting gate has no significant effect on the input switching threshold level, or on the '0' state output level. However, leakage through the OFF expander phase-splitter transistors can increase the voltage drop across R2, which results in a lowering of the '1' level output voltage. The output voltage will be reduced by about 1 V to 1·6 V/mA leakage current through the expander terminals. Normally the leakage current is very small (less than 0·01 mA) so this effect is unlikely to cause serious embarrassment in machine design.

The leakage current in the OFF state also results in the semiconductor capacitances in the output stage being partially charged, so this leakage current has the same effect on the high frequency transfer characteristic of the gate as a shift in the d.c. baseline. (See section 9.4.5.) This loss in a.c. noise immunity will be partially offset by the extra capacitance at the expander points which must be charged (or discharged) by the noise before the gate output can respond.

Thus the d.c. '0' level noise margin is unaffected, but the d.c. '1' level noise margin and the high frequency noise immunity are decreased slightly by the addition of 'OR' expanders to an inverting gate.

10.5.4.2 'OR' *Expanders on non-inverting gates*

'OR' expansion of non-inverting gates does not affect the input switching thresholds or the output voltage levels, so d.c. noise margins are not affected. High frequency noise immunity is increased by the extra capacitance which must be charged or discharged before the gate can respond.

10.5.4.3 'AND' *Expanders on eight-input gate*

'AND' expansion of an eight-input gate does not affect the input switching threshold or the output voltage levels. As in the case of the non-inverting expander, high frequency noise immunity is increased by the extra capacitance present.

The 'AND' expander points are much more susceptible to noise than the two types of 'OR' expansion. In particular, the collector connection is at a very high impedance point on the base of the phase-splitter. However, as stated in Section 7.4, this should not affect normal working if the connections are kept short.

10.5.5 EFFECTS OF EXPANDERS ON DISSIPATION

The dissipation of an expanded gate will be very slightly increased by the leakage currents flowing through the resistors in the gate package. This effect will add less than a tenth of a milliwatt to the normal dissipation, so it can be ignored.

10.5.6 EFFECTS OF EXPANDERS ON THE SWITCHING SPIKE

The effects on the supply rail current spike of adding 'OR' expanders to an inverting gate are mainly capacitive, and are described in Section 10.2.6. The addition of 'AND' expanders to the 180 Ex does not affect the supply rail current spike, and neither does the addition of 'OR' expanders to a non-inverting gate.

11

Flip-flops

11.1 Types of Flip-flops

The four major types of flip-flop used in digital systems are the R.S., the M.S., the J.K., and the D-type.

11.1.1 R.S. FLIP-FLOPS

An R.S. flip-flop (reset–set) is formed when two multi-input 'NAND' gates each have one input fed from the output of the other gate (the simple bistable circuit). Earthing a free input to one gate 'sets' the flip-flop; earthing a free input to the other gate 'resets' it.

Generally R.S. flip-flops are built up from basic gate elements as required.

11.1.2 M.S. FLIP-FLOPS

The M.S. flip-flop (master–slave) can be made up by cascading two R.S. flip-flops with transfer gates between the outputs of the 'master' flip-flop and the inputs to the slave. If input gates are added to the 'master' flip-flop, and these input gates are coupled to the transfer gates so that one set is enabled while the other is inhibited, information can be set into the master, then later transferred through to the slave after the signals on the input lines have changed. Simultaneously clocked M.S. flip-flops can interchange data without any risk of malfunctioning, because each master is first set to the state of the slave in the other flip-flop, then the slaves are set to their own masters. Direct set and clear lines to the slave can be provided, and multiple input gating to the master can be used so that an M.S. flip-flop can be a powerful and valuable logic element. 'Two-sided' data inputs may be needed for correct operation of an M.S. flip-flop, i.e. 'data' signals are fed to one side of the master, and 'not data' to the other side, and the operation of the flip-flop is determined by the enabling or inhibiting of the input and transfer gates from a clock line. 'Data' (Q) and 'not data' ($\bar{Q}$) outputs are usually available.

11.1.3 J.K. FLIP-FLOPS

The distinguishing feature of any J.K. flip-flop is that if both the 'data' (J) inputs and the 'not data' (K) inputs are held high and a clock signal is fed to the device, the output state will be complemented. Otherwise the action is similar to that of an M.S. flip-flop. The complementing action can be achieved by the provision of internal links from the Q output to a K input and from the $\bar{Q}$ output to a J input on an M.S. flip-flop. Thus when all other inputs are held 'high', the inputs will 'see' the complemented output of the flip-flop, since the Q output normally follows the logical state of the J input. Multiple inputs can be used, and direct set or direct reset can be applied in the same way as for an M.S. flip-flop.

11.1.4 D-TYPE FLIP-FLOPS

A D-type flip-flop (delay) merely transfers the data present on its input, prior to a rising clock edge, to the output after the clock edge. Generally, D-type flip-flops do not have a 'complemented' input, but complemented outputs are usually available. A complex D-type flip-flop is described in detail in Chapter 13.

11.1.5 TYPES AVAILABLE IN T.T.L.

All types of flip-flop are available in T.T.L. (see manufacturers' data sheets). J.K.s are the most common, and are available with single J and K inputs, with two flip-flops to a package, up to complex versions with 'AND-OR' input gating, or with J, not-J, K and not-K inputs. Flip-flops with not-J and not-K inputs are especially useful when only uncomplemented data are available, as without the complemented input facility extra gates would have to be used as inverters. S.U.H.L., Series 54/74, and Series 9000 have each adopted different logical configurations for their flip-flops.

11.2 Working of Flip-flops

Each of the major T.T.L. families has adopted a different mode of operation for flip-flops. When flip-flop circuits or more complex elements are being considered, it must be remembered that logic and circuit diagrams published in data sheets are usually only approximate representations of the circuits actually implemented on the chips. The planar process permits the use of circuitry that could not be properly built up from discrete components. For instance, resistive areas can be built into base diffusions, and diodes can be included to limit the swing of transistor collectors.

11.2.1 DIRECT COUPLED FLIP-FLOPS

There are several types of direct coupled flip-flops in use. In almost all cases the circuits look complex and difficult to follow, but this is often because of multiple J and K inputs and direct clear and set lines. The easiest way to understand any such complex device is to go through the circuit and strip it of all complications such as direct clear or set lines, and then study the fundamental flip-flop action.

11.2.1.1 *Direct coupled master–slave J.K. flip-flops*

The simplest type of T.T.L. J.K. flip-flop is the direct coupled master–slave which forms the basis of the 9000 Series flip-flops. Shorn of all trimmings this reduces to the basic circuit shown in Fig. 11.1. The back

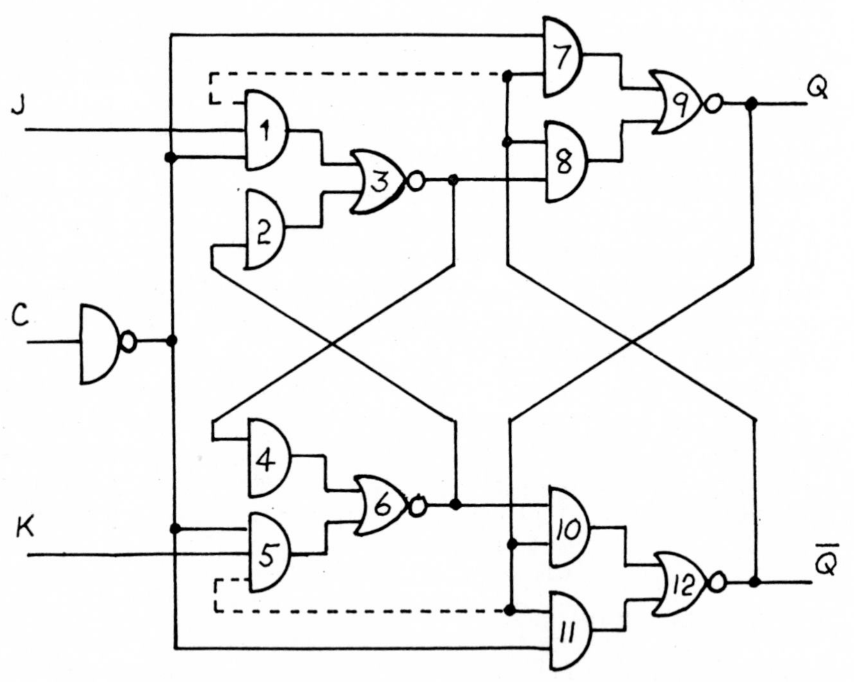

Fig. 11.1 Direct-coupled master–slave J.K. flip-flop.

connections to the J and K input gates are shown by broken lines. The circuit comprises two cascaded flip-flops, each of which is formed by a pair of 'AND-OR-INVERT' gates. Each side of each flip-flop has a gate which is enabled or inhibited by the internal clock line, and all four gates are enabled or inhibited simultaneously. When the internal clock line is high, gates 1 and 5 are enabled, and if complementary data are present on the inputs, the master flip-flop will set to the data. The high internal clock line also enables gates 7 and 11, which 'locks' the state of the slave flip-flop. As the internal clock line falls to a low state, gates 1 and 5 are inhibited but the information in the master bistable is 'held' by gates 2 and 4. Gates 7 and 11 are inhibited as the clock line falls so the data in the master bistable will be transferred to the slave via gates 8 and 10. If this transfer does not require the outputs of 9 and 12 to change state, there will be no disturbance to the output, but if a change of state is required, then the output which was low will go high, and then as this 'high' ANDs with the 'high' from the master to gate 8 or 10, the other output will fall to its low state.

Because this type of circuit imposes a loading of four T.T.L. gates on the internal clock line, an input buffer stage is provided so that the external clock-line loading is that of a normal T.T.L. gate. Thus the outputs of the flip-flop change when the external clock line is changed from a low to a high state, and the master flip-flop is enabled only when the external clock line is 'low'.

While the external clock line is 'low', a spurious positive going pulse on the 'low' data input can cause false setting of the device. Consider the case when input 'J' is low and 'K' is high, with output 'Q' low and '$\bar{Q}$' high (i.e. the output should not change when the clock pulse is applied). Since output 'Q' is low, the master bistable will be set with gate 6 low and gate 3 high. While the external clock line is 'low', the setting of the master bistable will remain undisturbed, because gates 1 and 5 each have two inputs high and one input low.

If the input 'J' goes 'high' momentarily, the output of gate 1 will go 'high', so gate 3, and hence gate 4 will go 'low'. Since the output of gate 5 is already 'low' (held low by the low input from gate 9), gate 6 will go 'high'. Gate 2 will follow gate 6, and then when the 'J' input returns 'low' the output of gate 3 is held 'low' by the 'high' input from 2, and the master bistable is 'locked' in a 'false' state.

Thus whenever a flip-flop of this type is used, care must be taken to avoid spurious signals on the input data lines, and the duration of the 'low' external clock pulse should be as short as possible. The manufacturers' data sheets quote only typical clock-pulse widths, so unless the user's purchasing specification defines an unambiguous limit, some safety margin must be left when selecting a clock-pulse width.

Direct set inputs (when present) are fed into gates 4, 7, and 8, and direct clear inputs are fed into gates 2, 10, and 11. If one of these direct inputs is taken 'low' while the external clock line is 'high', both master and slave bistables are 'forced' to the desired condition.

11.2.1.2 *Transistor coupled master–slave J.K. flip-flops*

The basic master–slave flip-flop of the Series 54/74 family is similar to that of the 9000 family, except that instead of feeding the internal clock line direct to the input gates of the slave flip-flop, it is taken to the emitters of two coupling transistors (9 and 10 on Fig. 11.2). When the clock line is 'high', these transistors isolate the slave flip-flop from the master.

When the clock line changes from 'high' to 'low', transistors 9 and 10 turn on and 'lock' bistable 7 and 8, and the data is transferred to the slave bistable 11 and 12, while simultaneously the master bistable input gates 1 and 5 are inhibited.

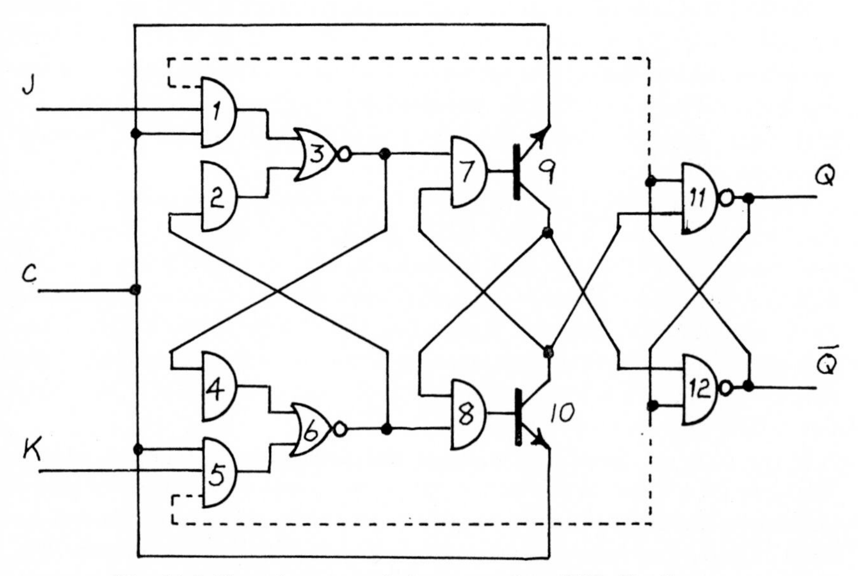

Fig. 11.2 Transistor-coupled master–slave J.K. flip-flop.

These flip-flops operate from a positive going clock pulse (since there is no internal buffer on the clock line), and the clock input loading is equal to that of two normal gate inputs. When the clock line is 'high', the flip-flop can be falsely set by noise on a 'low' input in the same way as the direct coupled flip-flop. Minimum clock-pulse widths are defined as 20 ns.

Direct set inputs are fed into gates 4, 7, and 11, and direct clear to gates 2, 8, and 12.

11.2.1.3 *Edge-triggered J.K. flip-flops*

The noise susceptibility of the master–slave flip-flops is overcome in the edge-triggered flip-flops which are available in the Series 54/74 range. Each side of these flip-flops consists of two 'locking ring' circuits, and there are two internal bistables (see Fig. 11.3). The clock line enables or inhibits the second gate on each side. Thus when the clock line is 'low', the input gates 1 and 4 are isolated from the rest of the circuit. When the clock goes 'high', gates 2 and 5 are enabled, and the 'rings' 1 and 2 (and 4 and 5) store the input data. Similarly, the second 'rings' 2 and 3 (and 5 and 6) are enabled. If both data inputs are held 'low' while the clock input changes from low to high, feedback from bistable 3, 6 to bistable 2, 5 will hold the output state unchanged. If both data inputs are held 'high', the output will be complemented as the clock input changes. Bistable 3, 6 ensures that the outputs are not affected when the clock input changes from high to low.

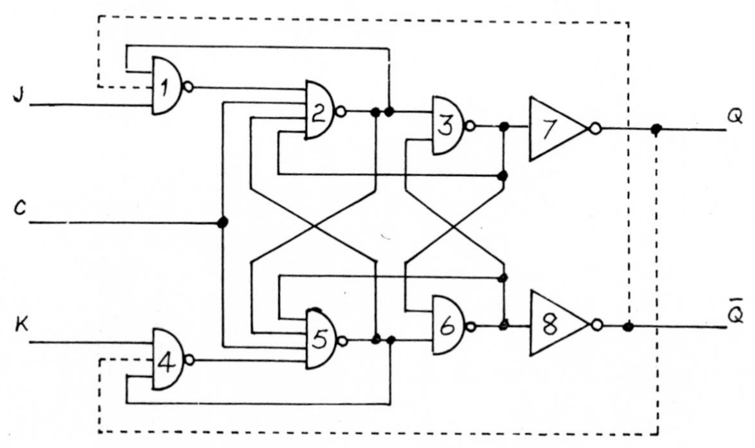

Fig. 11.3 Edge-triggered J.K. flip-flop.

Direct set inputs are applied to gates 4 and 6, and direct clear to 1 and 3. These flip-flops are capable of slightly faster operation than the master–slave types.

Edge-triggered flip-flops are not affected by false input levels while the clock is in its low or high steady state, but the input data must be steady at its correct level for a period (called the pre-set time) before the active edge of the clock occurs, and the data may have to remain constant for a further period (post-set or hold time) after the active clock edge. Any noise on the data inputs during the pre-set or post-set periods may cause malfunctioning.

11.2.1.4 *Edge-triggered D-type flip-flops*

The Series 54/74 edge-triggered D-type flip-flop consists of three cascaded bistables, as shown in Fig. 11.4. A low input on the clock line isolates the output bistable from the input gate by inhibiting gates 2 and 4. When the clock input rises from low to high, the input data is staticized in the three bistables, and thereafter any change of state of the data input will be 'blocked' either by a 'low' from gate 2 into gate 1, or by a 'low' from gate 4 into gates 2 and 3.

Direct set inputs are applied to gates 3 and 5, and direct clear to gates 1, 4, and 6.

The S.U.H.L. edge-triggered D-type flip-flop is explained in detail in Chapter 13.

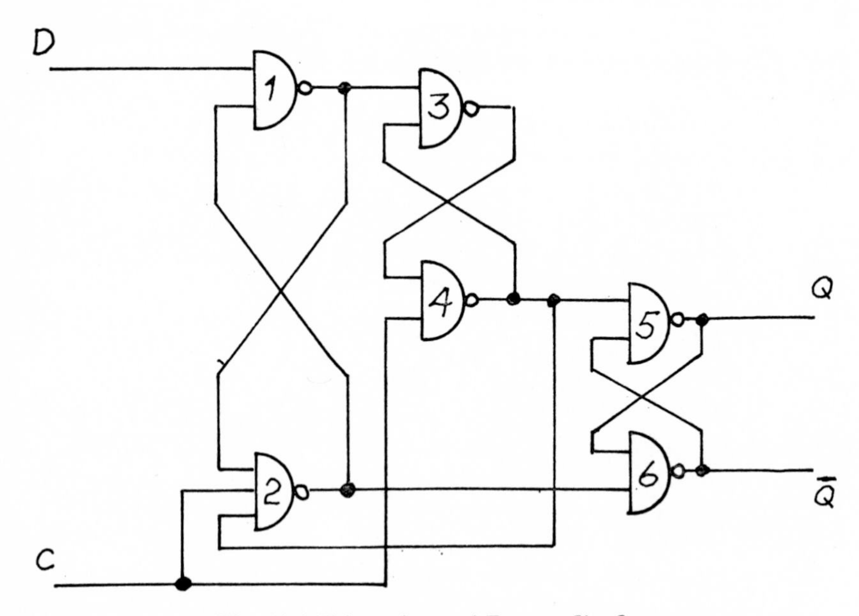

Fig. 11.4 Edge-triggered D-type flip-flop.

11.2.2 A.C. COUPLED (CHARGE STORAGE) FLIP-FLOPS

The charge storage flip-flop (used in the S.U.H.L. range) has similarities to the transistor coupled flip-flop described in Section 11.2.1.2, in that the clock input directly inhibits the data input gates and is also connected to the emitters of the transfer transistors. However, the resemblance ends there.

The output stage is a normal Series 2 bistable, but its operation is made faster because the collectors of the transfer transistors (VT3 and VT6) are fed directly to the bases of the phase-splitters (see Fig. 11.5).

The bases of the transfer transistors are fed from the emitters of transistors which correspond to the phase-splitters in the input gates (VT2 and VT5). They are also connected to the collectors of the input bistable transistors, VT7 and VT9.

While the clock is 'low', the input gates are inhibited, and the bases of VT3 and VT6 will be held at earth by R3, R4, R5, and R6. When the clock input rises (while complementary data are present on the inputs) either VT1 or VT4 will be enabled, and VT2 or VT5 will receive base current. The current through VT2 or VT5 will flow to the bistable network. If VT1 is enabled, the current through VT2 flows to earth through R6, and R5 in parallel with the base-emitter of VT8. VT8 is thus turned

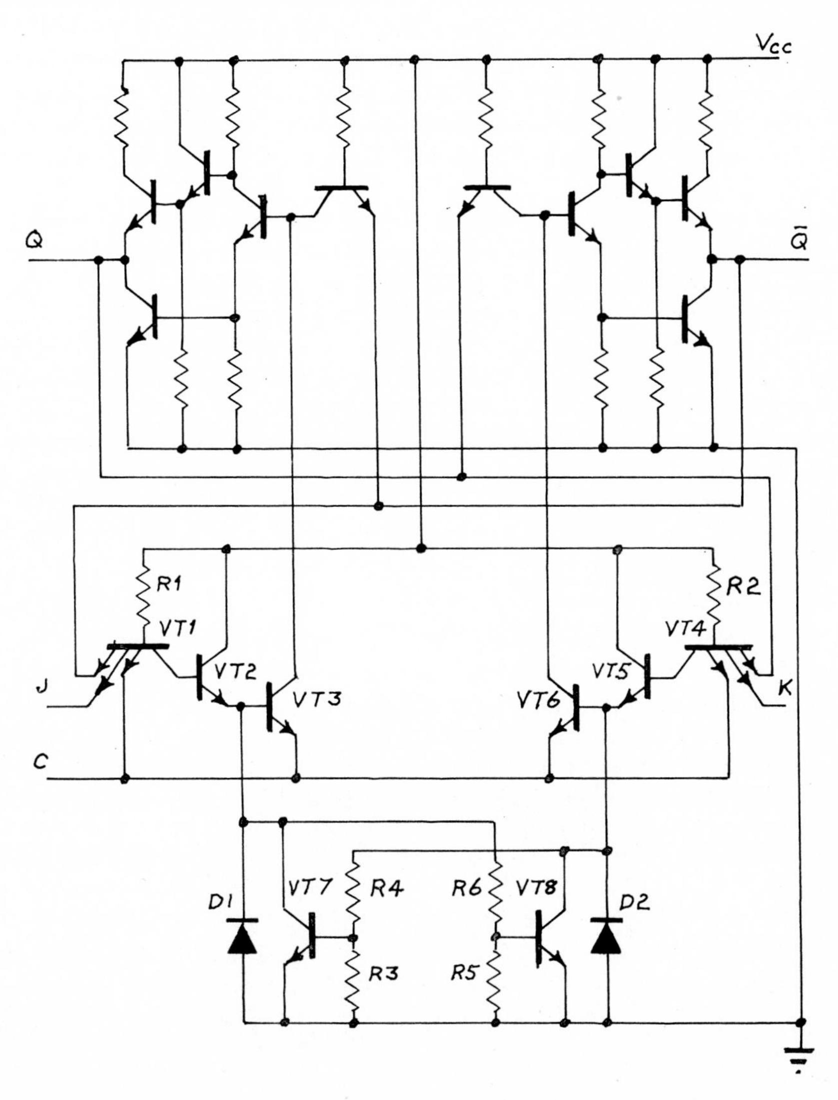

Fig. 11.5 A.C. coupled (charge-storage) J.K. flip–flop—basic circuit.

on, so the base of VT7 is held 'low'. No current can flow through VT3 or VT6 because the emitters are held 'high' by the clock line.

If a spurious signal on the input of VT4 turns VT4 and VT5 on, the current through VT5 will flow to earth through VT8, and the setting of the bistable will be undisturbed.

When the clock line falls, current flow through VT2 (or VT5) ceases. The state of the bistable VT7, VT8 is maintained by the stored charge in the depletion region of diode D1 (or D2). This stored charge is sufficient to provide base drive for VT3 (or VT6), which sets the output bistable to follow the state in which the inputs were when the clock input went high. The stored charge leaks away after the output bistable has been set.

The charge stored in the diode will enable the bistable to hold its state if the enabled input gate is turned off by a short noise pulse, while the clock input is high.

These charge storage flip-flops are fast in operation, and are 'dual-edge-triggered'; a rising clock edge sets the input bistable, and a falling clock edge sets the output bistable. At all other times, the bistable has full noise immunity regardless of the state of the clock input.

The input bistable can be reset while the clock line is high if both J and K inputs are held low until the stored charge has been dissipated through R3 and R4 or R5 and R6. If the flip-flop is given a full clock pulse with the J and K inputs held low, the output state will remain unchanged.

11.3 Use of Flip-flops

11.3.1 INPUT LOADINGS

Whenever integrated flip-flops are used, the data sheets should be read carefully. Data inputs are usually standard T.T.L. circuits, but 'clock', 'set', and 'reset' inputs may have different characteristics from a normal gate input. The input circuits might comprise two normal T.T.L. emitter inputs, or there may be other non-standard circuitry so that the number of 'clock', 'set', or 'reset' inputs which can be driven by a standard gate will be less than the normal fan-out of the gate.

11.3.2 UNUSED HALF PACKAGES

When 'dual' flip-flop packages are used, there may be occasions on which half a package is left unused. Such an unused flip-flop should be 'locked' in its lowest dissipation state. (See appropriate manufacturer's data sheet.)

11.3.3 CLOCK EDGE SPEEDS

All T.T.L. flip-flops checked by the author (by no means all types available) were found to be capable of responding to clock edges which are, by T.T.L. standards, slow (in the microsecond range). However, it must be borne in mind that normal T.T.L. edge speeds are very fast, and flip-flops are designed to be controlled by T.T.L. devices. Attempts to use really slow clock edges, such as transformed down 50 Hz mains, will usually cause malfunctioning. Generally, clock pulses with edges slower than 100 ns should not be used. Although slower edges may not cause malfunctioning, slight variations in the clock threshold voltages may cause unacceptable clock skew in a register or counter if the clock pulse edges are not kept sharp.

12

Medium-scale Integration

12.1 What is M.S.I.?

M.S.I. devices are large, complex logic functions which are manufactured in the same way as T.T.L. gates, and are normally packaged in larger versions of the 14-pin packs used for the standard gate functions (or in standard 14-pin packages).

It is difficult to define a precise limit at which a device should be called an M.S.I. device and not merely an integrated circuit. One set of limits which has been suggested is that an M.S.I. device should contain between 25 and 100 gate functions. Some firms regard the quad exclusive-OR (true exclusive-OR) package as a gate package; others regard it as M.S.I.

M.S.I. devices are not usually made from an array of complete T.T.L. gate circuits interconnected by the metallization. Output stages are provided only where they are needed on the actual output terminals; or where a high internal fan-out is required: phase-splitter OR stages may have their emitters connected to earth through a diode to maintain the correct threshold level; and flip-flops may comprise only a pair of multi-emitter transistors. In many devices a whole set of flip-flops may be controlled by having their emitters connected to an internal clock line so that the set can be enabled or inhibited in the same manner as the transfer transistors in some master–slave flip-flops (see Section 11.2.1.2).

12.2 Advantages and Disadvantages of M.S.I.

The advantages of M.S.I. devices are the same as the advantages of integrated circuits—reduced package count, which means lower initial costs; fewer interconnections and hence improved reliability; shorter interconnections, with consequent reductions in cross-talk and random noise pick-up; lower overall system dissipation, which means less cost in power supply units; and quicker turn-around in system and board design.

There are disadvantages to be set against these advantages. The cost of an M.S.I. package is higher than that of a standard gate package, and the value of boards which carry significant numbers of M.S.I. devices will be much more than the value of similar boards carrying normal gate packages. Wastage of complete boards must therefore be eliminated if costs are to be cut by the use of M.S.I. devices.

The dissipation of an M.S.I. device can be up to ten times the dissipation of a standard gate package, although the size of a 16-pin M.S.I. package is almost the same as that of the 14-pin gate package. If a group of such high dissipation M.S.I. devices is mounted all in one area of a board and no special cooling is provided, devices in the middle of the group could run at very high temperatures. In one instance a worst-case chip temperature of over 250°C was calculated for an M.S.I. counter.

Good thermal design becomes critical when significant numbers of M.S.I. devices are used on densely packed boards, and if the thermal design is inadequate, there is a very real risk that the final equipment may be less reliable than a larger version built up from standard gate packages, instead of being more reliable.

Another point which must be given careful consideration is the input loading on M.S.I. devices. A 'clock' or 'reset' line might drive a number of circuits on the chip, and the input currents can be several times those of a normal T.T.L. gate input.

12.3 M.S.I. Families

At the time of writing, both the 54/74 and the 9000 ranges offer M.S.I. devices. In the 54/74 family, the numbers of the M.S.I. devices are a continuation of the gate and flip-flop numbers, and they are actually mixed so there is no number above which it is obvious that the devices are M.S.I. and not ordinary integrated circuits (e.g. 7474 and 7476 are flip-flops, but 7475 and 7477 are M.S.I. latches). The 9000 family M.S.I. devices are all numbered in the 9300 series.

In the 54/74 M.S.I. range there are a few 54H/74H devices, and a number of 54L/74L types available, as well as the standard speed devices. At the time of writing, Fairchild have announced a 9200 range, which has lower dissipation (and speed) than the 9300 range.

So far as is known to the author, all the 54/74 M.S.I. devices follow normal Series 54/74 standard practice on all inputs and outputs (except for special 'open collector' driver outputs).

The 9300 series devices do not all follow normal 9000 series practice. On the data sheets for some devices the output stage is shown as following 54/74 practice (i.e. a diode on the emitter of a single 'pull-up' transistor), whereas for some other devices the two transistor 9000 series output stage is shown. For the remaining devices in the series the data sheets do not specify which type of output stage is used. Although all devices (presumably) conform to similar specification requirements, the different output

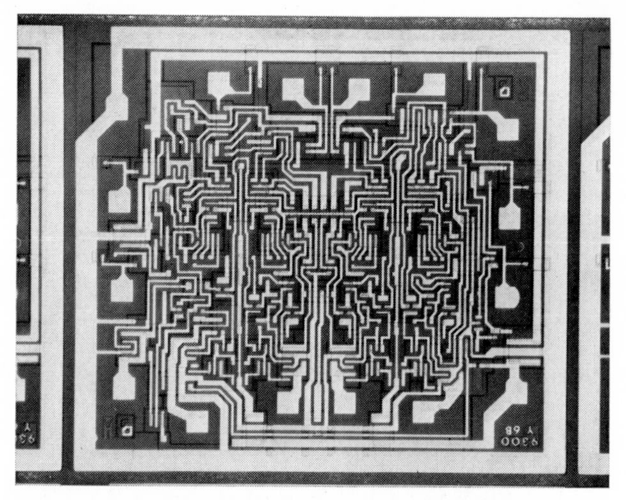

Fig. 12.1 M.S.I. Chip layout.
Fig. 12.1 (Courtesy of G.E.C. Semiconductors Ltd.)

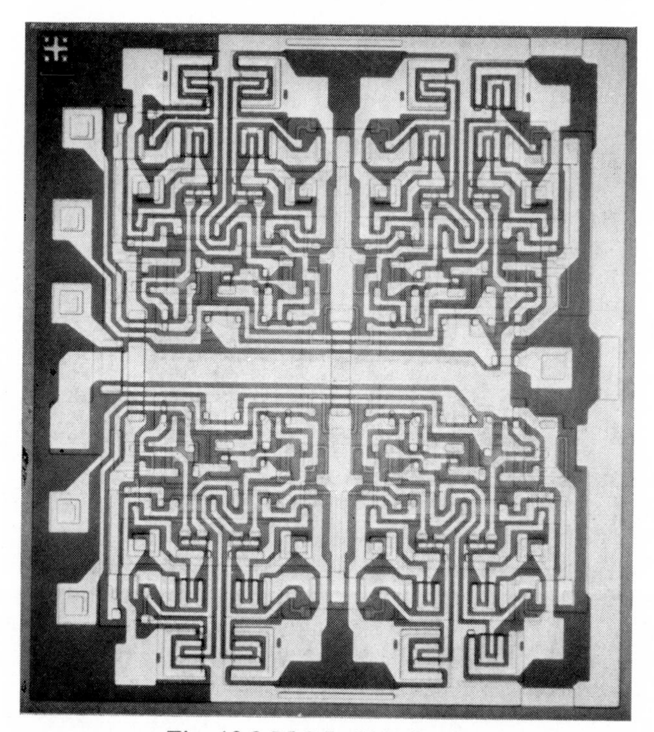

Fig. 12.2 M.S.I. Chip layout.
Fig. 12.2 (Courtesy of National Semiconductor Corp.)

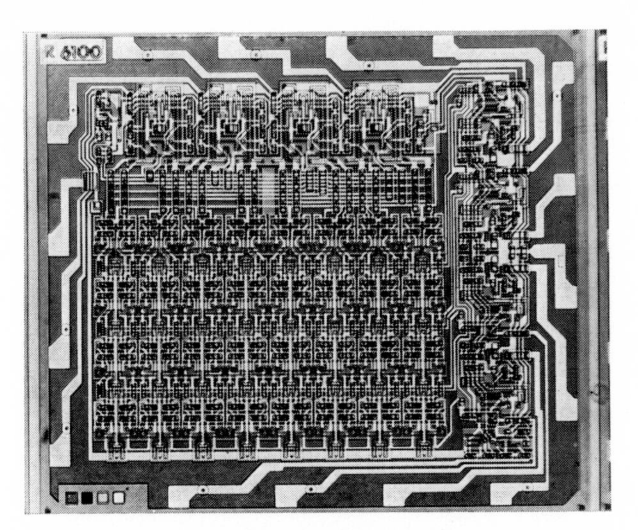

Fig. 12.3 M.S.I. Chip layout.
Fig. 12.3 (Courtesy of Raytheon Company.)

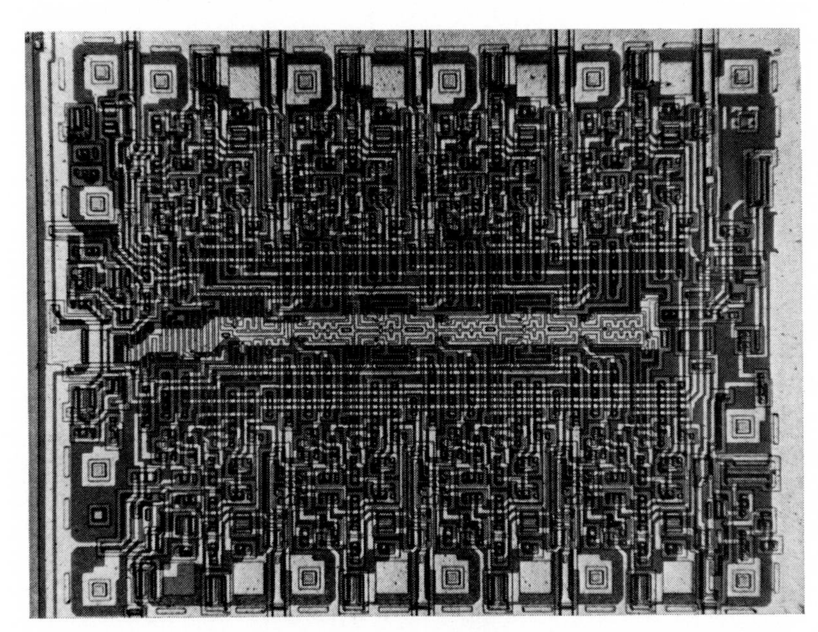

Fig. 12.4 M.S.I. Chip layout.
Fig. 12.4 (Courtesy of Texas Instruments Ltd.)

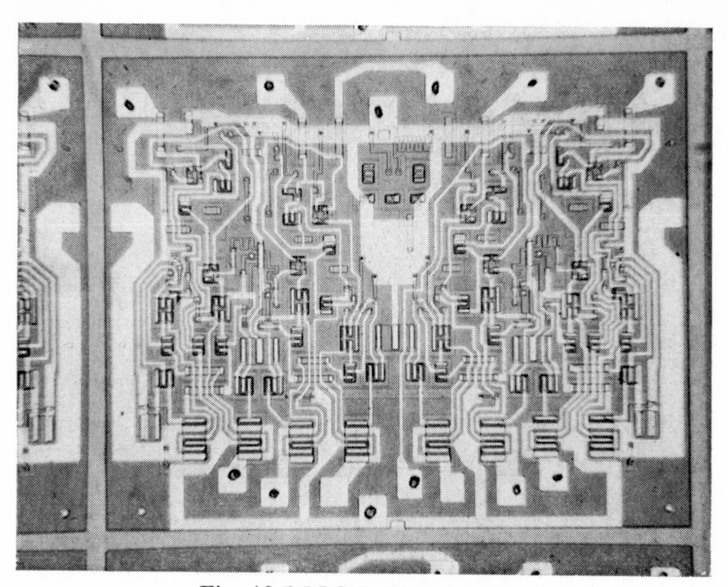

Fig. 12.5 M.S.I. Chip layout.

Fig. 12.5 (Courtesy of Transitron Electronic Ltd.)

stages used can result in different turn-off performances when lines of critical length are driven (see Chapter 14). As edge speeds are not quoted in the data sheets, the determination of critical line length is difficult.

At the time of writing, most if not all, makers of 9300 series devices also make Series 54/74 M.S.I. devices. The converse does not apply—several firms make 54/74 M.S.I. but do not make Series 9300. The two families originated at the peak of the boom in microelectronics; whether both families will survive the present recession in the industry is questionable. The 54/74 range of M.S.I. is much larger than the 9300 range and it is possible that the 9300 range may be superseded by extensions to the 54/74 range.

12.4 Range of M.S.I. Devices

As has already been mentioned the future of the two major families of M.S.I. is uncertain. Also, the future of some devices in the range is uncertain. Devices have been designed and advertised, but not yet been put into production, and there can be no certainty that all the devices currently listed in the various makers' catalogues will still be listed in five or six years' time. On the other hand, it is quite likely that new M.S.I. devices will be designed, will find a ready market, and will be available 'off the shelf' at competitive prices.

Because of these uncertainties, only the present trend in M.S.I. types can be described. Any list of functions available will almost certainly be out of date by the time this book is printed.

The main groups of functions currently available comprise registers, latches, encoders and decoders, counters, adders and arithmetic units, multiplexers, comparators, and memory elements.

12.4.1 REGISTERS

One of the commonest M.S.I. registers is a serial-in, serial-out, eight bit shift register, a simple array of eight master–slave flip-flops with a common clock line and an input gate. Each flip-flop sets to the data of the preceding stage on the clock pulse, and complemented data are available only from the outputs of the eighth flip-flop.

Another common register is the four bit parallel-in, parallel-out, shift-left register. The parallel data inputs can be externally connected to the data outputs to provide serial-in, parallel-out, shift-left, shift-right working, and, with suitable external gating, full parallel input, shift-left, shift-right working can be achieved. Such devices can be used as shift and storage registers, serial to parallel converters, or parallel to serial converters.

12.4.2 LATCHES

The four, six, and eight bit latches available are simply groups of bistables, usually with a common 'clock' or 'set' line, which work on a single data input to each bistable.

12.4.3 ENCODERS, DECODERS, AND MULTIPLEXERS

The commonest M.S.I. decoder is a four line binary to decimal decoder-driver which has ordinary single transistor output stages on its ten output lines. Probably the next most common type is the B.C.D. to seven-segment decoder which is designed to drive 'two square' seven line solid state indicator devices.

Other decoder types cater for other coded inputs such as Gray code.

Multiplexers select the data on one of two, four, eight, or sixteen input lines, the selection being controlled by up to four (binary) control lines.

These types are all complex 'AND-OR' devices, which may include 'strobe' or other common control functions.

12.4.4 COUNTERS

Counters available in M.S.I. include such types as four bit up-down counters, hexidecimal, decade, and divide by twelve counters. All types comprise a group of cascaded J.K. flip-flops (not the full 'discrete' flip-flop circuits described in Chapter 11) with feed back connections to achieve the desired count. Some of the 9300 series counters have 'carry' gating such that a number of devices can be cascaded to provide fast wide-range counters.

12.4.5 ARITHMETIC UNITS, ADDERS, AND COMPARATORS

The 9300 range offers a full four bit arithmetic unit which performs the functions 'add', 'subtract', 'AND', and 'exclusive OR' (not equivalent). The other arithmetic devices available are plain full adder circuits which require external gating to enable them to perform full arithmetic functions, but which can be used in other applications besides arithmetic units.

The comparators available examine two sets of four bit data lines and have three outputs for A greater than, equal to, and less than B. Three extra input connections enable these devices to be cascaded with only one gate delay for each four bits added. Eight bit parity generators can also be cascaded in a similar manner.

12.4.6 MEMORIES

One of the earliest M.S.I. devices available was the 16 bit read–write memory element. Read-only memories are available with 256 bit capacity, for which the stored data is specified by the customer when the device is ordered. As these large read-only memories are partially 'custom built' they can never be supplied 'off the shelf'.

It is in the field of memories that M.O.S. devices are most likely to supersede T.T.L., as devices such as a 1000 bit read–write memory element would be quite impracticable in T.T.L. technology.

13

Detailed Consideration of a Flip-flop Circuit

This detailed consideration of a single device has been included as an example of the lengths to which it may be necessary to go to determine the suitability of a particular device for use in a high-reliability equipment. The device considered is the S.U.H.L. or H.L.T.T.L. multi-input D-type flip-flop, but similar studies might be applied to any M.S.I. or other complex device.

A 'black-box' approach to the use of integrated circuits may result in cheap initial designs, but all too often savings made in not gaining a full understanding of the devices before designs are begun can be lost many times over if any troubles occur during machine commissioning.

13.1 Introduction

The S.U.H.L. or H.L.T.T.L. D-type flip-flop (1D.F.F.) is a 50 MHz binary storage device with single-sided input connections in positive logic, and normal (Q) and complemented ($\bar{Q}$) outputs. There are eight data input connections arranged in two three-way and one two-way 'AND' gates, 'ORed' together, and a 'hold' input which enables the device to retain a logic '1' output on Q regardless of the data set on the normal inputs. (See Fig. 13.1.)

Input data is staticized and stored, and the outputs reset, only by a rising edge on the clock input. The input data must be constant for a short time before the clock edge rises (pre-set time) and also for up to 2 ns after the rising clock edge (post-set time). At all other times the output state is unaffected by any changes on the data or hold inputs, and a falling clock edge is ignored. The outputs settle to their new levels 5–20 ns after the rising clock edge, the '0' output settling before the '1' output. The small difference between the maximum post-set time and the minimum settling time means that when D.F.F.s are directly interconnected (e.g. as in a shift register) care must be taken to ensure that all D.F.F.s will receive the clock signals virtually simultaneously (within 3 ns) or, if this not possible, delays must be inserted between the output of one D.F.F. and the input of the next to accommodate the clock skew.

All data and hold inputs and the outputs are similar to conventional T.T.L. circuits, but the clock input is a unique circuit which sinks its load current in the '1' state, and may sink a small current in the '0' state, instead of sourcing current in the '0' state and sinking only leakage current in the '1' state. The clock input also has a higher threshold level than a standard T.T.L. gate. For correct operation the clock input pulse must have edges faster than 50 ns. Special care must be taken to avoid introducing noise on to a D.F.F. clock line driven by a T.T.L. gate at its

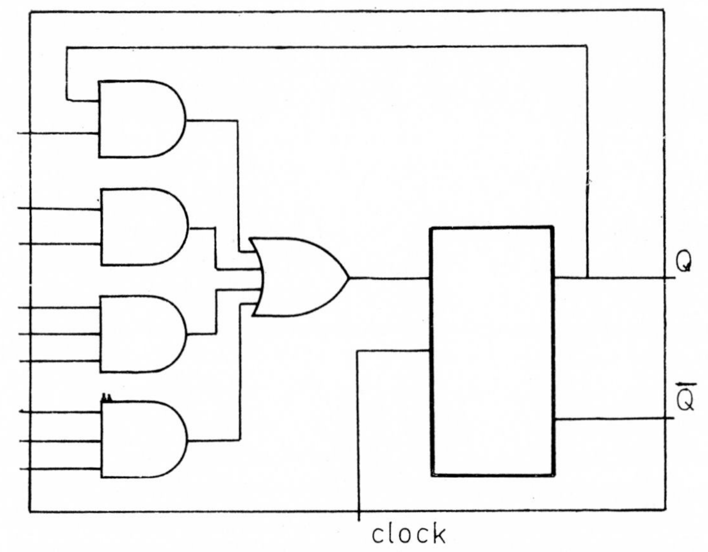

Fig. 13.1 The S.U.H.L./H.L.T.T.L. D-type flip-flop.

'1' level. The higher threshold means that '0' level noise can be ignored.

As the action of the device is dynamic and for most of the time the logical state of the inputs is ignored, a conventional truth table is of little value, and some of those published are dangerously misleading. The action of the device is most simply described by the following statements.

'After a rising clock edge the "Q" output sets to the logical sum of the data present on the inputs immediately before and during that rising clock edge.

'A falling clock edge is ignored.

'A logic "1" on the hold input will allow the "Q" output to change from a "0" to a "1" in response to the input data present on the rising clock edge but will prevent the Q output from changing from "1" to "0".

'The $\bar{Q}$ output is the logical inverse of the Q output.

'The logical sum of the input data is given by

$$D = D_1.D_2.D_3. + D_5.D_6D_7. + D_{13}.D_{14}.'$$

Normal working is similar to that of an M.S. flip-flop, except for the limited period in which the device accepts input signals and that setting of the output is initiated by the same clock edge as the staticization of the input data. J.K.-type operation, i.e. a change in output state for every clock pulse can be achieved by connecting the $\bar{Q}$ output to one of the input gates (usually enabled by a 'count' signal) and inhibiting the other input gates. The D.F.F. can be used for building parallel entry shift registers, multi-input registers, pulse train counters or parallel entry addition or subtraction counters, all with the minimum of external gating. The immunity of the device to spurious input signals at all times other than over the rising clock edge makes it especially suitable for use in registers where the input data has a prolonged settling period or where the input data may be subject to a high level of noise both before and after setting the register.

13.2 Logical Operation

13.2.1 GENERAL

A schematic circuit diagram of the device is shown in Fig. 13.2, and an equivalent logic diagram in Fig. 13.3. The circuit diagram is broken into blocks to correspond with the lettered logic elements on Fig. 13.3. Dotted convention is used on the logic diagram to emphasize the points at which inversions occur.

13.2.2 LOGICAL STATE WITH LOW CLOCK INPUT

When the clock input is low, element e has its output high, (logical '0'), enabling elements a, b, d, h, m, and n. Incoming data are 'ANDed' in elements a, b and d and the results 'ORed' in element g, the output of which can be called 'D'. (i.e. the logical sum of the input data). (The operation of the hold element h is dealt with in Section 13.2.6.) Element j inverts D. Since the output of element c is at '0', element f is inhibited and its output is at '0', so it has no influence on the data sum D at g.

Elements m, n, r, and s form the output bistable. Elements k and q are held inhibited by element c so their outputs are at '0' and cannot affect the setting of the output bistable. Elements l and p are uninhibited, and their outputs can be affected by changes in D (or $\bar{D}$) from j or g. It requires a spurious '1' from l or p to upset the output bistable, and this cannot occur because l and p are also connected into the bistable. For instance, l can give a '1' output only if the $\bar{Q}$ output is '1'. This '1' from

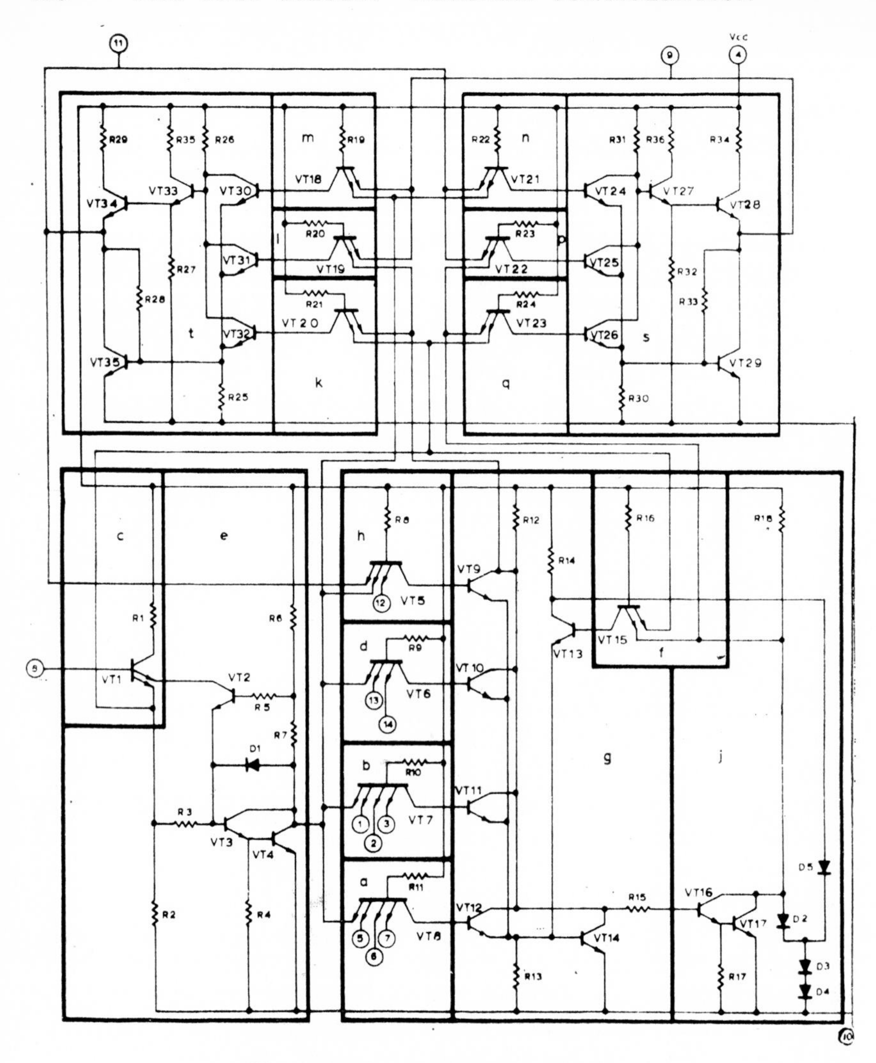

Fig. 13.2 Circuit of D-type flip-flop.

element l would set Q to '0'. But for $\bar{Q}$ to be '1', Q must already be '0'. Figure 13.4 shows the D.F.F. with a low clock input, with predetermined logic levels indicated.

13.2.3 LOGICAL STATE WITH HIGH CLOCK INPUT

When the clock input is high, the output of element e will be low, so elements a, b, d, h, m, and n will be inhibited, with their outputs at '0'.

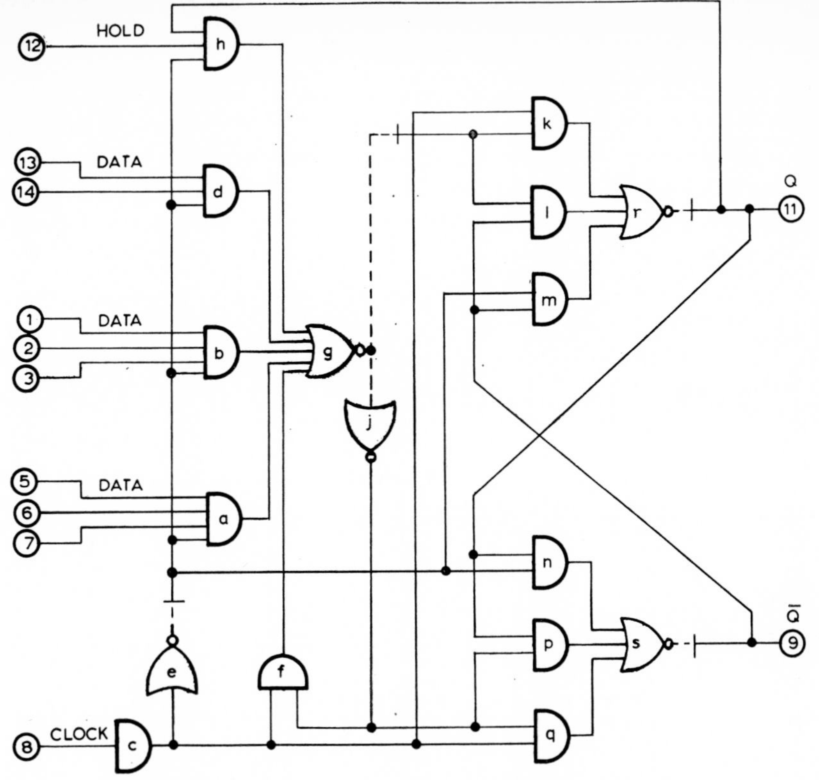

Fig. 13.3 Equivalent logic diagram of D-type flip-flop.

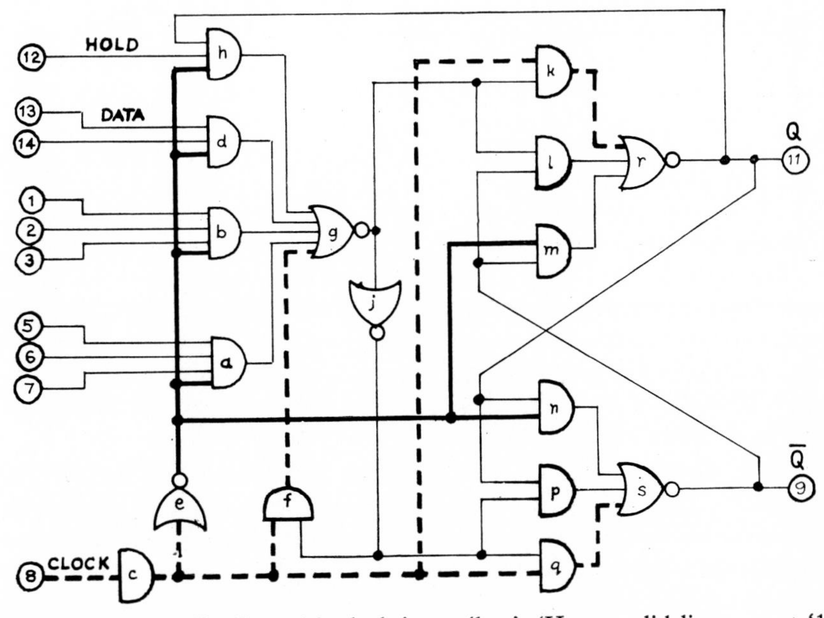

Fig. 13.4 D-type flip-flop with clock input 'low'. (Heavy solid lines are at '1' level, dotted at '0'. All other lines according to data present.)

Element f will be enabled by the high output of element c, so f, g, and j form a closed ring to which all inputs are held at '0', so that whatever the state of the ring, it cannot be changed while the clock input is high. Elements k and q are enabled by element c, so that D on j appears on the output of q and $\bar{D}$ (D from g inverted) appears at k. Whichever of these is

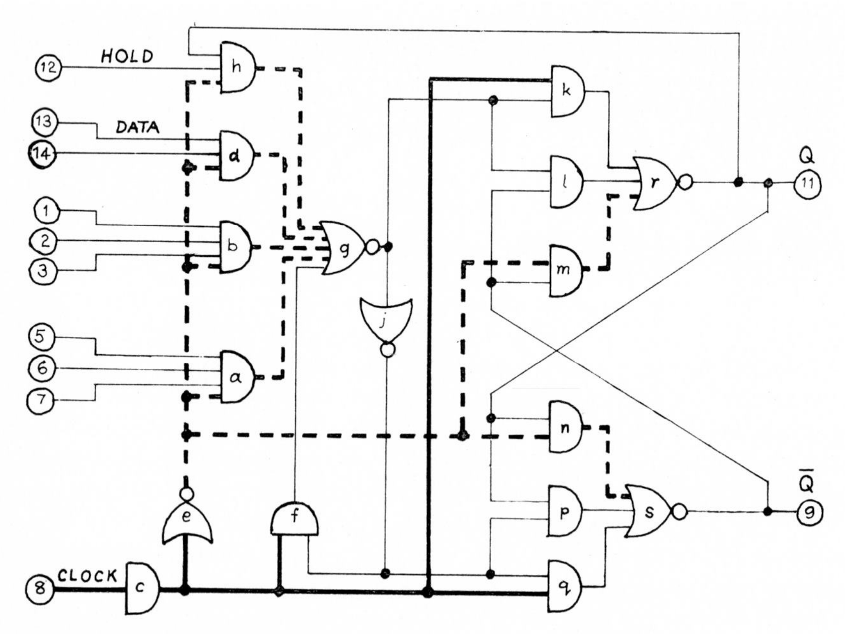

Fig. 13.5 D-type flip-flop with clock input 'high'. (Heavy solid lines are at '1' level, dotted at '0'. All other lines according to data present.)

'1' will set r or s, and the output bistable will be held via elements l and p. (See Fig. 13.5.)

Thus with the clock input high the output bistable will be set to the logical state of the inner ring f, g, and j, and since all inputs to this ring are inhibited, the bistable setting cannot be changed.

13.2.4 LOGICAL OPERATION WITH RISING CLOCK INPUT

While the clock input was low, changing input data could change D on elements g and j. For correct operation the input data must not change during the pre-set or post-set periods. This ensures that as the clock input rises D will be steady. As the clock input rises, the output of element c rises with only a slight delay. Elements f, k, and q are enabled, again with only a small delay. The enabling of element f locks D into the ring f, g, j; and the enabling of k and q means that a new '1' from D can reset element r or s, giving the new '0' output on Q or $\bar{Q}$, which feeds through l or p to set the new '1' output.

The output of element e falls after the clock input has risen, inhibiting all the input elements, and also inhibiting m and n.

The logical operation is thus such that on the rising clock edge the ring of elements f, g, and j is closed a fixed time before the input elements are inhibited and any environmental conditions or device degradations which would tend to slow the closing of element f will also slow down elements a, b, d, and h, which will have similar delays to f, so the device is always safe from logic race problems. It could well be dangerous to use this mode of operation with a discrete component circuit, or even a circuit built up from separate gate elements.

13.2.5 LOGICAL OPERATION WITH FALLING CLOCK INPUT

The falling clock input first inhibits elements f, k, and q, setting their outputs to '0'. Elements m and n have their outputs at '0' initially, but l or p must have a '1' due to the output bistable action, and this '1' will hold the bistable set. The '0' from element f will propagate through elements g and j, but before this '0' could upset element p, element e will have gone high following the falling clock input, enabling elements m and n and holding the output bistable still locked. The rising output of element e also enables the input elements a, b, d, and h.

13.2.6 LOGICAL OPERATION OF HOLD ELEMENT

The hold element h has a single external input (pin 12), its other two inputs coming from the clock generator element e and the Q output, element r. When the input pin 12 has a high level input and the clock input is low, element h will have the same logical output as the Q output. If this is '1', the D.F.F. will remain set to '1' on the next rising clock edge since *g* will give a '1' output if any of its inputs are '1'. If the Q output is '0', then element h will be '0' and the D.F.F. will set to the logical sum of the data in the normal manner. Thus enabling the 'hold' input will allow the D.F.F. to remain at '0' or change to '1' under the control of the input data (and the clock) but once the D.F.F. has set to a '1' this will be retained until the D.F.F. is clocked with the hold input inhibited (and '0' on the data inputs).

13.3 Electrical Operation

13.3.1 GENERAL

It can be noted that all the elements except c, e, g, and j are similar to conventional T.T.L. gate elements. Study of the chip layout of a D.F.F. shows minor differences from the published circuit diagrams (R35, R36, and D5 are absent from all other known published circuits) which might not apply to all devices. The working of the actual circuits studied is described here. Any variations will clearly be minor and will have little significant effect on the working.

13.3.2 OPERATION OF THE CONVENTIONAL ELEMENTS

Elements a, b, d, f, h, k, l, m, n, p, and q are all multi-emitter 'AND' gates, in which a '0' on any emitter causes the collector to fall to a '0' and in which all emitters have to be '1' for the collector to be '1'.

Elements s and r are 'NOR' gates similar to the standard S.U.H.L. 2 T.T.L. gate output stage. If the base of any of the three input transistors is at a '1' then that transistor will saturate and the output will be a '0'. A '1' output is obtainable only if all three inputs are '0'. The only differences between these elements and a conventional T.T.L. gate output stage are resistors R28 and R33. These reduce the reverse bias on the bases of the output transistors VT29 and VT35 and keep VT28 and VT34 active with no external load when at logic '1' output. This tends to increase switching speed and attenuate spurious '1' level ringing.

13.3.3 ELECTRICAL OPERATION OF ELEMENTS c AND e (the clock generator)

13.3.3.1 *General*

Electrically these elements are one circuit with two outputs, but the logical working is easier to appreciate if they are drawn separately. These elements, which together generate the clock signals controlling the operation of the device, are relatively complex, but an understanding of their working is essential to appreciate the overall device operation. An easy way of understanding the operation is to make some assumptions, then if these prove to be false, to make fresh assumptions until the correct conditions are reached.

Figures 13.6 *et seq.*, show these circuit elements, with the current flow paths emphasized. Nominal resistor values were quoted by Bhola of Transitron Ltd. A V_{CC} of 5·0 V, saturated base-emitter potential drops of 0·8 V, saturated collector-emitter potential drops of 0·3 V, forward-biased diode potential drops of 0·7 V, and common emitter current gains of 10 are assumed throughout.

13.3.3.2 *Input low*

Assumption 1. Assume the input is at 0 V, and that transistors VT1, VT2 VT3, and VT4 are all OFF. The only current which could flow is that shown in Fig. 13.6, which would result in the potentials shown. The base

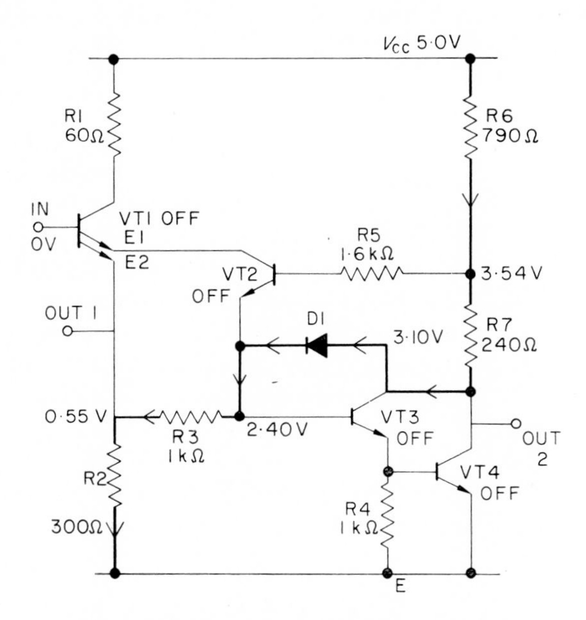

Fig. 13.6 Clock generator. Assumption 1.

of VT3 could not be at 2·40 V while VT3 is off, so this assumption is false.

Assumption 2. Assume that the input is at 0 V and that transistors VT1 and VT2 are OFF but that VT3 and VT4 are ON. The base potential of VT3 must be 1·6 V as shown in Fig. 13.7. Resistors R2 and R3 must be biased to 1·6 V and this requires 1·22 mA which can come only through D1, which must therefore be forward biased. Hence neither VT3 nor VT4 can saturate and the output 2 potential will be 2·3 V as shown in Fig. 13.7.

A current of 2·82 mA flows through R6 and R7. 1·22 mA of this must flow through R2 and R3 to bias VT3 and VT4 ON, which leaves 1·60 mA to flow into the collectors of VT3 and VT4. Of this, 0·8 mA must flow from the emitter of VT3 into R4 to bias VT4 ON. Therefore 0·8 mA remains to flow into the collector of VT4, which will require a base drive of 0·08 mA, as will VT3. These base drives are sufficiently small not to upset the assumed bias.

The emitter of VT2 is at 1·6 V and the potential between R6 and R7 is 2·8 V as shown. Therefore 0·38 mA of base current into VT2 can flow via R5, so this assumption must also be false.

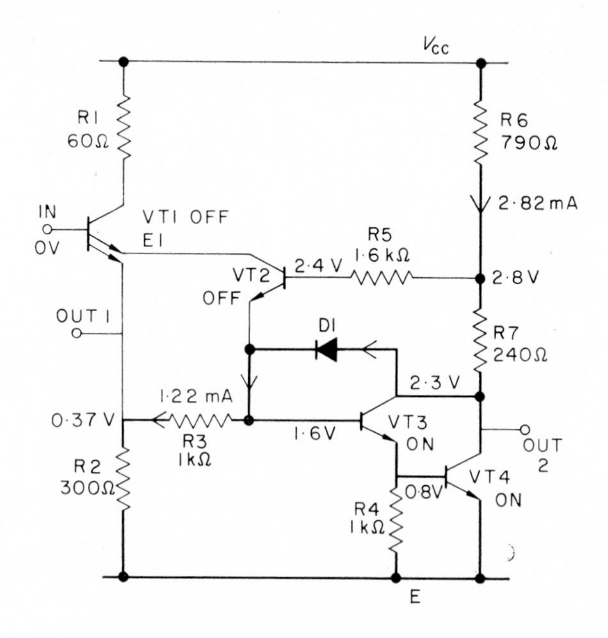

Fig. 13.7 Clock generator. Assumption 2.

Assumption 3. Assume that the input is at 0 V and that transistor VT1 is OFF; VT2 is ON but starved of collector current and VT3 and VT4 are ON.

This state is similar to that in the second assumption except that now current is shared between D1 and the base-emitter diode of VT2. The collector of VT2 cannot conduct until the emitter '1' of VT1 is at least 0·1 V above the emitter of VT2 (i.e. 1·7 V), which would reverse bias VT1 and therefore only negligible leakage current can flow. There are now no contradictions, so the assumption can be taken to be correct.

Assumption 4. The preceding assumptions have ignored the connection of output '1' of the clock generator to the rest of the D.F.F. circuit. This point is connected to the emitter of VT15, VT20, and VT23, all of which are two-emitter transistors, in which the voltages on the other emitters depend on the logical sum of the input data. When the data sum is '1' the second emitter of VT20 will be held down at 0·3 V by VT14, and the

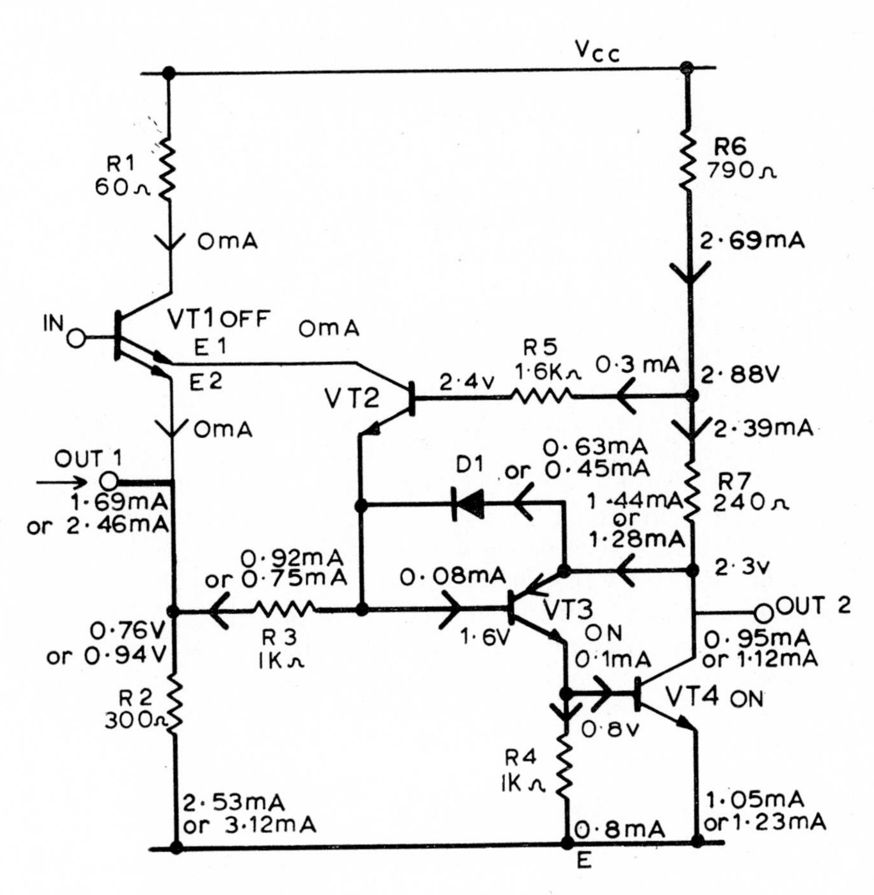

Fig. 13.8 Clock generator. Stable state with input '0'.

second emitters of VT15 and VT23 will be high; when the data sum is '0' the second emitter of VT20 will be high and those of VT15 and VT23 will be held at 1·1 V by VT16 and VT17.

When the clock input is low, output '1' of the clock generator cannot be below 0·37 V, so if the data sum is '1' VT14 will take all the current through R21 and VT20, and the clock generator will sink current from R16 and VT15 and R24 and VT23. If the data sum is '0', calculations show that output '1' of the clock generator will be at 0·94 V, so the clock generator will continue to sink the current from VT15 and R16, cutting off the second emitter of VT15 which is at 1·1 V, and the clock generator will also sink the current from R21 and VT20. Thus output '1' of the clock generator must always sink the current from either two or three T.T.L. input stages (nominally 4 kΩ resistors). Figure 13.8 shows the current and voltage levels in the clock generator circuit with the input low. Where pairs of figures are shown, the upper figure refers to an input data sum of '1' and the lower figure to a sum of '0'.

With the clock input at less than 0·7 V, output '1' of the clock generator ('c' on Fig. 13.3) will be at 0·76 V or 0·94 V depending on the data input, and output '2' ('e' on Fig. 13.3) will be at 2·3 V with transistors VT3 and VT4 on but held out of saturation.

13.3.3.3 *Input high*

With the input high, VT1 can be ON and can supply base current to VT3 via R3. VT3 can saturate, so its emitter will sit at 0·8 V (base potential of VT4) and its collector at 1·1 V. Current flow through R6 and R7 will hold their junction at 2·0 V, but as this is only 0·4 V above the base of VT3 and hence the emitter of VT2, VT2 will be OFF (see Fig. 13.9). D1 is completely reverse biased.

Output 1 of the clock generator will be V_{BE} VT1 below the clock input (1·7 V or more) so VT15, VT20, and VT23 will have their emitters connected to output 1 cut off. Output 2 will be at 1·1 V with transistors VT3 and VT4 on, VT3 saturated. The current into output 2 can be that from up to five T.T.L. input stages depending on the states of the input data. This current into VT4 will have no significant effect on the output voltage.

13.3.3.4 *Input rising from low to high*

If we consider the input starting at any voltage less than 1·56 V (V_{BE} VT1 plus the voltage on R2 in Fig. 13.8) and rising slowly, nothing will happen until the input potential reaches 1·56 V. Then as the input passes 1·6 V, emitter E2 of VT1 will conduct, and the potential on output 1 will rise, keeping 0·8 V below the input. The current from emitter 2 of VT1 begins to flow through R2, raising the potential on the junctions of R2 and R3 until eventually the base drive for VT3 is coming through R3 instead of through D1 and the emitter-base junction of VT2. As the current flow through D1 ceases, VT3 can saturate, and its collector potential falls to 1·1 V.

During the transition, the collector of VT2 rises to a higher potential than its emitter while the potential on the base of VT2 is falling. The stored base charge in VT2 causes a burst of collector current which over-drives VT3 and causes it to turn on quickly. This action occurs as the input to VT1 passes through approximately 2·5 V, when the potentials immediately after switching are as shown in Fig. 13.9.

Further increases in input potential merely raise the potential at output 1.

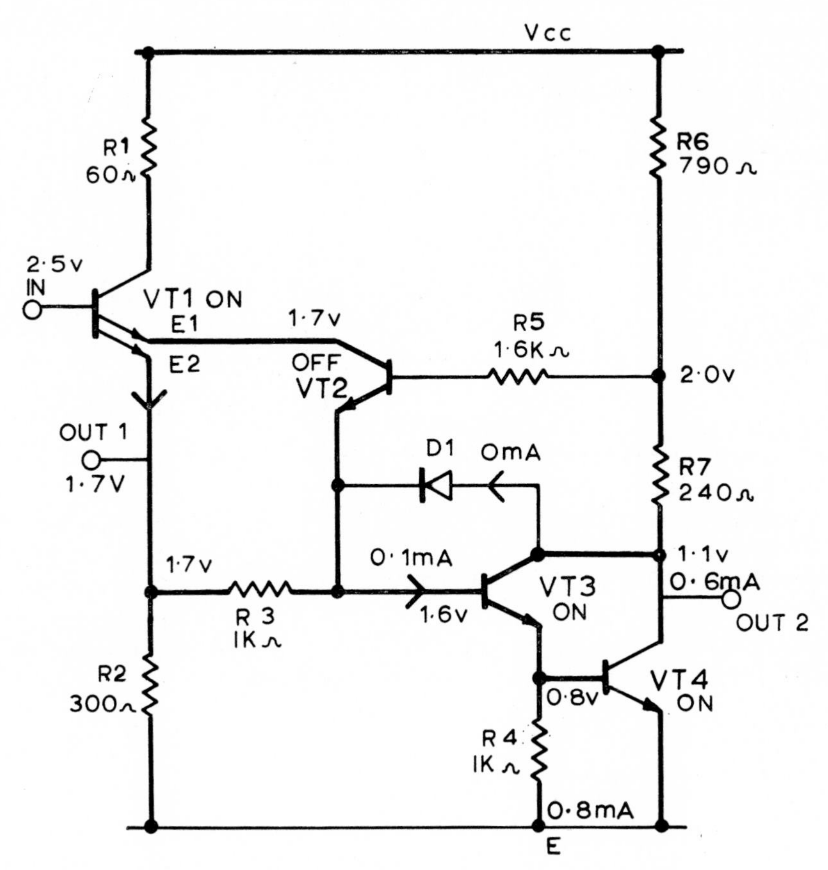

Fig. 13.9 Clock generator. Stable state with input '1'.

13.3.3.5 *Input falling from high to low*

The reverse of the above mechanism occurs with a falling clock input, except that there is no 'kick' from VT2 to encourage faster switching of VT3.

Figure 13.10 shows the input and the two output waveforms for the clock generator when the input rises and falls very slowly. At normal T.T.L. logic edge speeds, output 1 will follow the input, delayed by

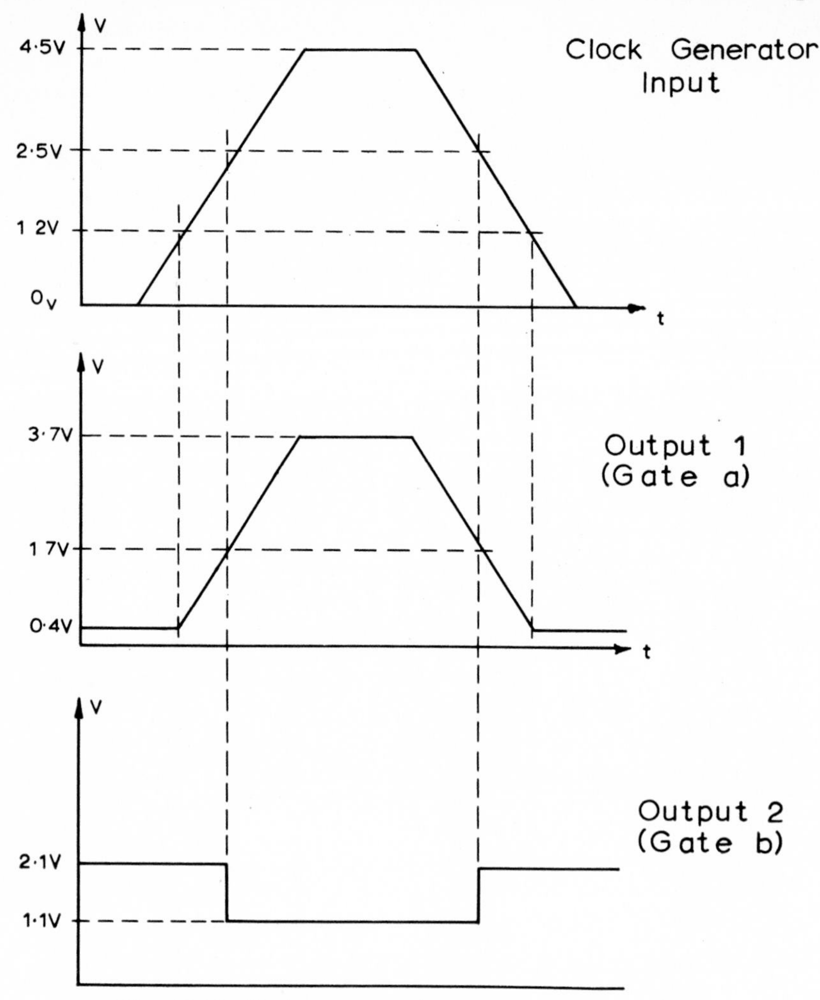

Fig. 13.10 Clock-generator input and output waveforms.

approximately 2 ns, and output 2 will switch about 6 ns after the input edge.

13.3.4 ELECTRICAL OPERATION OF ELEMENTS f, g, AND j (the Locking Ring) (See Figs. 13.11 and 13.12)

Element g is represented as a NOR gate although it is significantly different from elements s and r. Transistors VT9, 10, 11, and 12 are connected to generate the NOR of their base inputs on their common collectors at R12, but when VT9, 10, 11, or 12 is turned on by input element a, b, d, or the 'hold' element, VT14 is also turned on, its collector clamping to 0·3 V the collectors of VT9, 10, 11, and 12. When the inputs to VT9, 10, 11, and 12 are all low, these transistors and VT14 will be off, with their common collectors high.

Element j inverts the output of element g, the diodes D2, D3 and D4

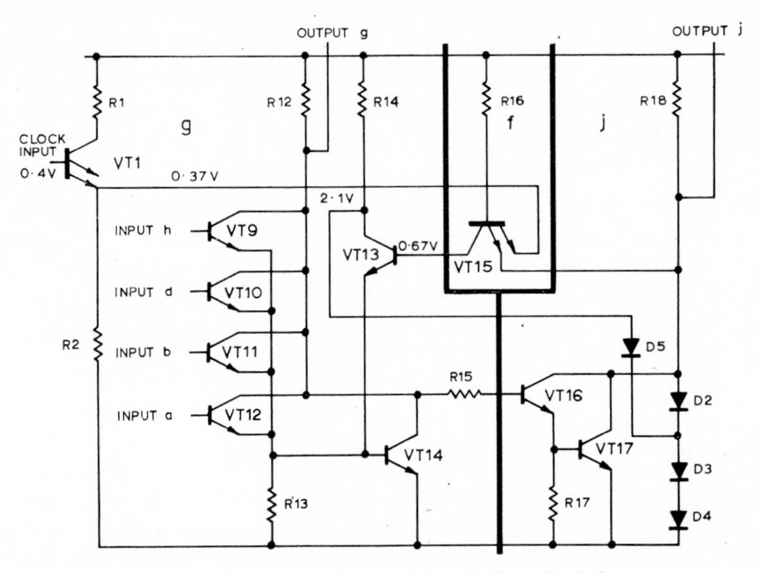

Fig. 13.11 Elements f, g and j with low clock input.

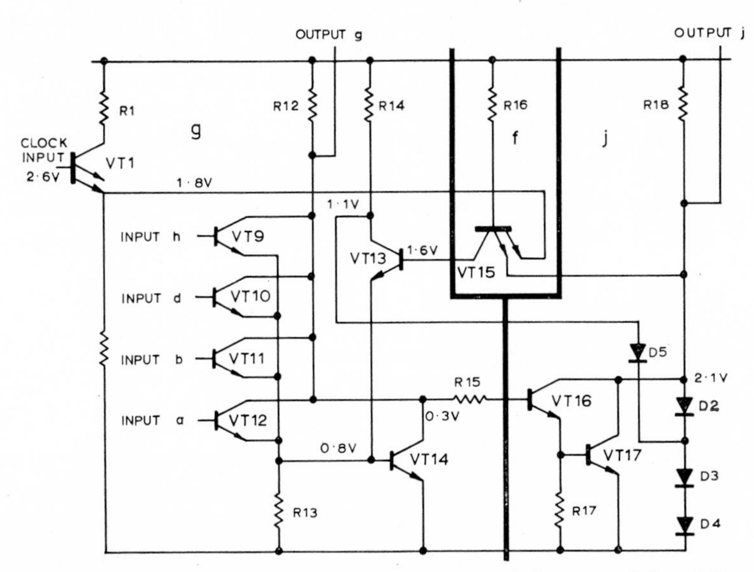

Fig. 13.12 Elements f, g and j with high clock input and data '1'.

limiting the output swing of element j to approximately 1 V, from 1·1 V (V_{BE}VT17 + V_{CE}VT16) to 2·1 V (D2 + D3 + D4). The output of element j feeds one input of element f, the other input of which is controlled by output 1 of the clock generator element c. Thus when the clock input is low the output of element c will also be low, and transistor VT13 will be cut off with its base at 0·67 V (0·37 V on emitter 2 of VT1 + V_{CE}VT15). When element f is enabled by a rising clock input, a low output from element j will still hold VT13 cut off while a high output from j will turn on VT13. The collector swing of VT13 is limited to approximately 1·1 V (V_{BE}VT14 + C_{CE}VT13) to 2·1 V maximum by D5, D3, and D4. Once turned on, VT13 will supply base current to VT14 regardless of the condition of VT9, 10, 11, 12, thus holding elements g and j locked in the logical state they were in immediately before the clock input to element f started to rise. This 'ring' remains locked while the clock input is high. Relevant potentials for a low clock input are shown in Fig. 13.11, and those for a clock input which has gone high while the logical sum of the data inputs was '1' are shown in Fig. 13.12.

13.4 Logical Use of the Clock or Hold Inputs

In any equipment it is unlikely that every D.F.F. will be required to respond to its input data with every timing signal generated. The master timing signal can be logically gated with the D.F.F. input gate controls such that a clock signal is fed to a D.F.F. only when it is required to operate, or the master timing signal can be fed to all D.F.F.s as a clock signal and the 'hold' input enabled whenever the D.F.F. is not required to respond to the clock.

On equipment using more than a few D.F.F.s the former method is to be preferred; although it may give rise to clock skew problems it avoids the problems associated with the distribution and generation of a fast, high current, noise-free clock signal. The gate which generates the actual D.F.F. clock signal should be on the same board as, and in close proximity to the D.F.F.s.

13.5 Typical Electrical Characteristics of the D.F.F.

13.5.1 GENERAL

Apart from the clock input, all inputs and outputs are similar to standard T.T.L. gate elements, and similar characteristics apply. The current from the supply rail is considerably higher than for a standard gate package, and the dissipation is quoted as being typically 150 mW with the clock input low. Typically dissipation with the clock input high is about 200 mW.

The chip capacitance between the supply rail and earth should also be higher than for a standard T.T.L. gate. Since it is impossible for both outputs to switch simultaneously, the D.F.F. should not generate a larger rail current spike on switching than a normal Series 2 (fast) gate element.

Clock threshold levels are typically 1·5–2·45 V, with a current requirement of 0·4 mA at 3·0 V. Most devices checked would accept clock edges as slow as 1 μs but a few devices have been observed which failed to operate with edge speeds between 200 and 300 ns.

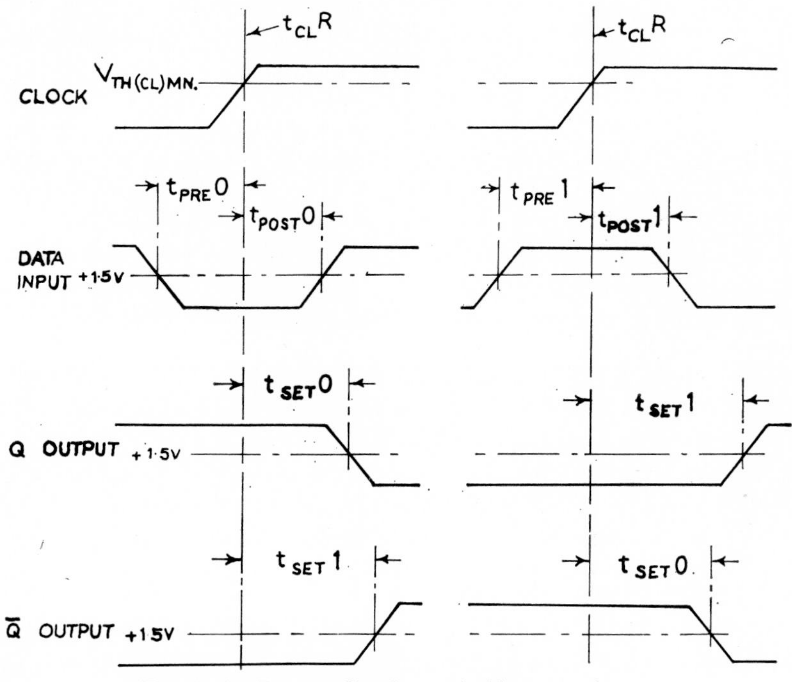

Fig. 13.13 D-type flip-flop switching waveforms.

13.5.2 TIMINGS (See Fig. 13.13)

Since it is the '0' level output of the bistable which sets the other side of the bistable to '1', '0' level settling times are about 4 ns faster than those for '1' levels on both Q and $\bar{Q}$. Timings for Q and $\bar{Q}$ are the same. Typically, a '0' level settling time is about 6 ns, with 10 ns for the 1 level. Pre-set times are always positive, although they may be as little as just over 1 ns. '0' level pre-set times are generally about 2 ns less than '1' level, with typical values of 2 ns for pre-set 0 and 4 ns for pre-set 1. Post-set times appear likely to be negative—i.e. the input data could be changed immediately before the clock input rises, and once again the '0' level value is less than the '1' level. Post-set '0' times are typically −2·5 ns, with post-set 1 typically zero.

13.6 Clock Line Driving and Noise Immunity

The specified '1' level input threshold of 2·6 V means that to guarantee 0·5 V noise margin the element driving the clock line must have a worst-case '1' level of 3·1 V. A standard T.T.L. gate will not satisfy this requirement as its output voltage can be 3·0 V at 0°C with a 5·0 V supply rail (S.U.H.L. 1 gate—other gates could be at lower voltages). Also, a worst-case low fan-out gate would not source enough current at that voltage to supply even one worst-case clock input.

Re-specification of the gate output to gurantee that it will source sufficient '1' level current to drive a D.F.F. clock input, and a gamble on noise margins, can be completely avoided by using a resistor up to the supply rail to supply the clock input current, and using the driving gate to short circuit this current to earth. The value of the resistor must be chosen such that with the supply rail at its highest toleranced level the driving gate can sink the current through the resistor. The number of clock inputs which can be fed from one gate is then determined by calculating the current available at a '1' level of 3·1 V.

For example, a 270 Ω resistor can be used with a S.U.H.L. 1 gate. A worst-case 5 per cent end of life resistor with a highest allowable rail of 5·5 V will require a gate saturating to 0·2 V to sink 20·6 mA; only marginally above the current a high fan-out slow gate is specified to sink at 0·45 V.

With a rail of 4·75 V, the resistor would supply 5·8 mA at 3·1 V, and the gate can source a further 1·2 mA, giving a total available current of 7·0 mA worst case—i.e. a fan-out of 10 D.F.F. clock inputs from one high fan-out gate.

Experiments have shown that a clock line driven as described above will not give a spurious clock signal from a '1' level at the lowest allowable rail voltage of 4·5 V when simultaneous pulses from two fast gates are sent along tracks parallel to and on 0·050 in centres either side of the clock line for a length of twelve inches.

If normal gate inputs are also to be driven from the clock line, the pull-up resistor must be increased in value, and the number of clock inputs which can be driven will be decreased.

13.7 Chip Layouts of the D-type Flip-flop

13.7.1 INTRODUCTION

As a glance at the circuit diagram would suggest, the D.F.F. chip layout is complex. Both Sylvania and Transitron chips have been photographed and the layouts studied. The two layouts are quite different. The

Sylvania chip (see Fig. 13.14) is rectangular, nearly twice as long as it is wide, whereas the Transitron chip is more nearly square. Sylvania use 'heart-shaped' multi-emitter transistors, whereas Transitron use rectangular or tee-shaped transistors. This does not appear to cause any difference in the leakage currents on the inputs. The circuit contains no heavy current (over 20 mA) nodes, and on both layouts the widths of the aluminium tracks are more than adequate for the currents they carry.

13.7.2 SYLVANIA LAYOUT

The Sylvania chip is about 0·095 in by 0·05 in. The layout is achieved with only twelve buried layer crossovers, two of which are in series with resistors. These crossovers are: between R29 and VT34; between D5 and D3; on the emitter of VT34; on the emitter of VT13; between the commoned emitters of VT18 and 21 and the clock generator; between the commoned emitters of VT 18 and 19 and the collector of VT29; between the commoned emitters of VT21 and 22 and the collector of VT35; on the commoned emitters of VT19 and 20; on the commoned emitters of VT22 and 23; between R34 and VT28; on the emitter of VT28; and between the emitter of VT8 and the clock generator. The crossovers in series with the resistors can be clearly ignored.

Two more of the crossovers are in positions on the output circuits where other manufacturers of T.T.L. type circuits place diodes and so can be ignored, one is in a diode chain; and all but one of the remainder are in series with the equivalent of a standard gate input. Thus it would seem that the only crossover at a point which might be at all critical is that on the emitter of VT13, where a drop of up to 0·2 V could be tolerated (see Fig. 13.11).

The most immediately obvious features of the layout are the input and hold circuit multi-emitter transistors VT5, 6, and 7 along one end of the chip (top, as drawn), then the input transistors to the bistable circuit, VT18, 19, 20, 21, 22, and 23 with their base resistors in the centre area of the chip. VT15 is also in this group; its connections appear confused because a track on the overlay crosses the transistor, without making contact, where the base contact could be expected. The base contact is in fact between the two emitter contacts. The power rail runs up the left-hand side, feeding various resistors, most conspicuous of which are R35 and R36.

The large, rectangular output transistors VT29 and VT35 are also obvious, and it is interesting to note that the earth line feeds through their emitter metallizing to reach the ends of the chip.

The track running round clear outside the output and earth pads is the clock generator feed to the emitters of the input elements.

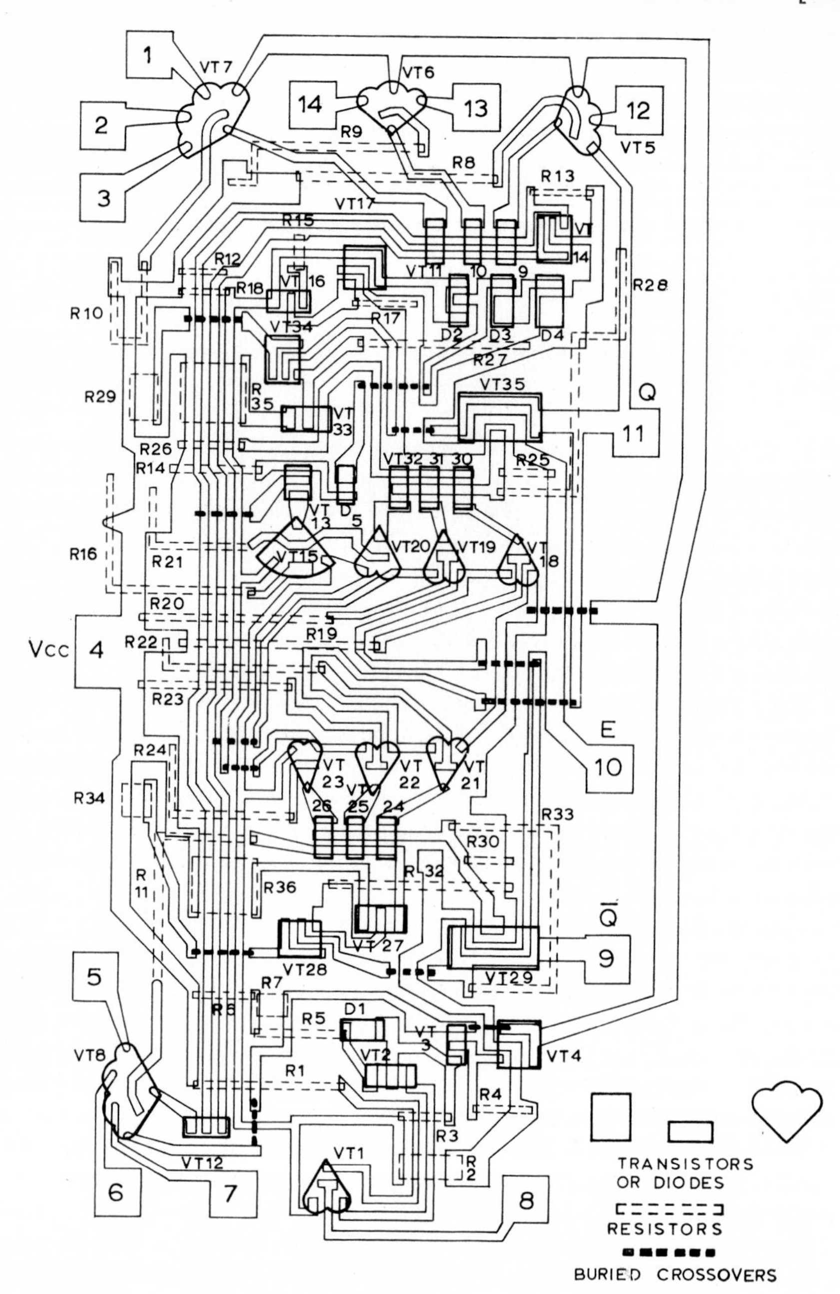

Fig. 13.14 Chip layout of D.F.F. (Sylvania).

It is also interesting to note that VT8 is well separated from the rest of the input gates, but has its inverting transistor VT12 immediately adjacent, with the long runs of track to join the rest of the circuit on what would be the 'OR' expander points of a standard gate.

The clock generator circuit occupies the lower right-hand corner of the chip, R2 being the most notable feature. The use of the common collector zone of VT3 and 4 as a buried crossover is liable to cause confusion when trying to follow the layout through (it is drawn here as a buried crossover). All diodes appear to have been diffused as normal geometry transistors with their collectors shorted to their emitters by the metallization.

13.7.3 TRANSITRON LAYOUT

The Transitron chip is about 0·075 in by 0·063 in, and the layout is more tightly packed than on the Sylvania chip (see Fig. 13.15). Six buried layer crossovers are used; one on the clock input from pin 8, one between the emitter of VT1 (clock generator) and VT20, and two in cascade on the connections from the Q and $\bar{Q}$ outputs back to the bistable inputs (VT21, 22, and VT18, 19). Thus all crossovers are in series with the equivalent of standard gate input circuits or the clock input, where the effects of the resistance of the crossover will be negligible.

The Transitron chip uses tee-shaped or rectangular multi-emitter transistors, and the diodes D3 and D4 are diffused as diodes, D4 having base-emitter diffusion with the base shorted to the collector by the metallization (i.e. a normal transistor diffusion as on the Sylvania chip) but D3 has no emitter diffusion, the metallization contacting the base diffusion directly. The diffusion of D1, D2, and D5 is obscure as these are all diffused within the collector areas of the transistors to whose collectors they are connected.

As on the Sylvania chip, VT8 is well separated from the other input transistors, but VT12 is in a common collector diffusion with VT5, 6, and 7, so the necessary long track is on the base of VT12 and not the collector and emitter as on the Sylvania chip. The length of track on the Transitron chip is considerably less, and the effect on the speed of working could be expected to be just about the same.

An obvious feature of the Transitron chip is the presence of unused crossovers and metallization windows (not shown on the drawing). These are used with other metallization masks to produce variants of the D.F.F. circuit.

The Transitron chip differs from the Sylvania in the Q and $\bar{Q}$ output stages, where R35 and R36 are omitted, and VT33 and VT34 have their collectors connected together, as do VT27 and VT28. There appears to be a facility present to include a low value resistance in these collector connections, the resistor normally being shorted out by the metallization.

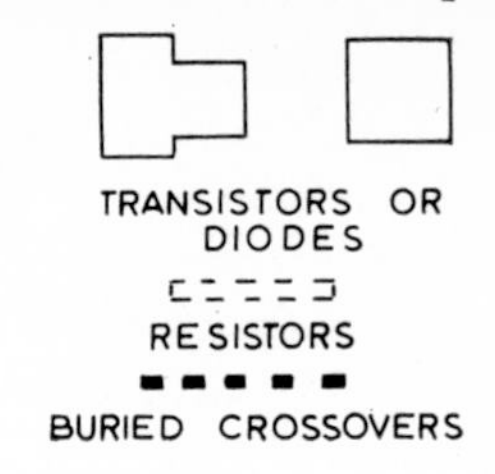

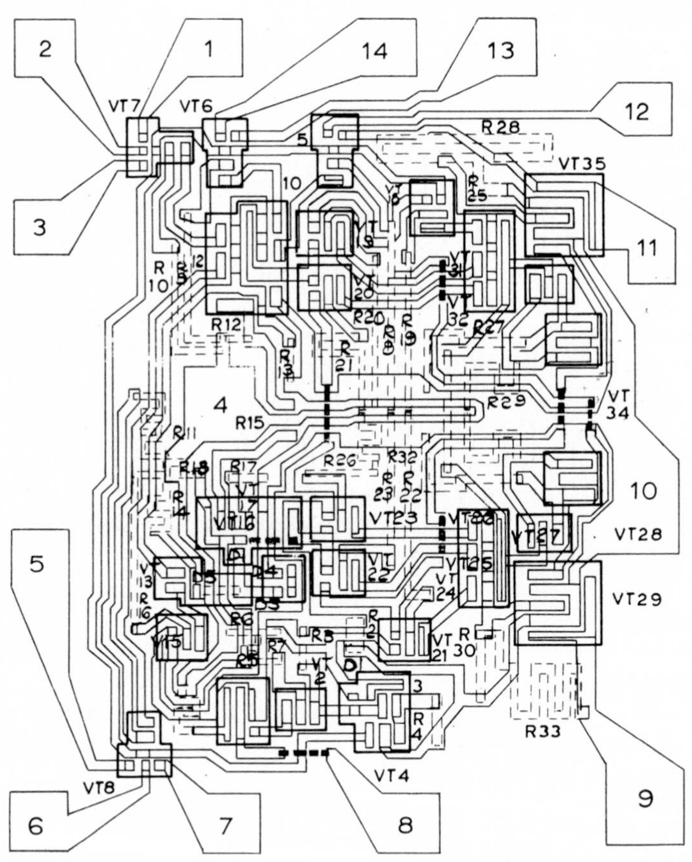

Fig. 13.15 Chip layout of D.F.F. (Transitron).

The distribution of capacitance over the layouts would appear to be such as to make even safer the correct operation of the device, which, if comparable sized transistors can be assumed to work at comparable speeds, is already quite safe from logic races.

From the chip layouts, it would appear that if any of the input gates of the 1.D.F.F. is not required logically, gate 'a' which uses pins 5, 6, and 7 should be the one left unused, or rather with its inputs earthed, unless both the 'used' gates require three inputs.

The fact that both layouts have lead bonding pads which are not right at the edge of the layout should not cause any risk of bond wires shorting as the normal D.I.P. mounting holds the chip at a much lower level than the lead frame and the wires rise steeply from the surface of the chip. Also the chips are oxidized after metallization, and the bond wires can contact their pads only through windows etched in this oxide layer.

Neither chip layout shows any feature which could be a potential hazard to long-term reliability. Input and output circuits are as well separated as they can be in a circuit of this complexity, and signal breakthrough should not occur.

14

Theory of Transmission Line Effects on Printed Circuit Boards

14.1 Introduction

Because digital logic systems depend for their operation on the voltage at a node being 'high' or 'low', i.e. at a logical '1' or '0', there may be a dangerous tendency to regard the operation of digital circuits as being almost a d.c. process, in which the only considerations needed in the design of printed circuit boards are to ensure that the necessary interconnection pattern is achieved, that all tracks are wide enough to carry the necessary d.c. current, and that the separation between tracks is adequate for electrical isolation and for the satisfactory manufacture of the boards.

In fact, this is very far from the truth. Inadequate design of gate interconnections can result in serious loss of speed in a digital equipment, and coupling of an unwanted signal into a gate input can result in the corruption of the data which is being handled. The transitions from level to level of the output of a T.T.L. gate occur in a very few nanoseconds, and if these fast edges are to be preserved, board designers must realize that they are designing equipment which must accommodate frequencies of hundreds of megahertz, and not just simple d.c. levels.

Although in practice designers can do much of their work without any need to calculate all parameters for every track on a board, users of T.T.L. should be familiar with the basic theory of transmission line effects.

If only one signal need be carried from one point to another, it would be simple to apply conventional transmission line theory to the problem. Unfortunately, the interconnection of integrated circuits involves complex networks of tracks or lines, and the pure theory becomes far too involved to be of any practical value in solving specific problems. This chapter and Chapter 15 describe the theory as used by the author and his colleagues in Computer Division, The Marconi Company, and describe the results of the practical tests done to extrapolate from the basic theory to suit the complex environment of a high speed data handling system. This section of the book is offered not as an exhaustive treatise on transmission lines, but as a practical guide to the user of T.T.L.

The two effects which most affect the use of T.T.L. are 'ringing' and 'cross-talk'. Ringing is the reflection of a pulse edge from an incorrectly terminated end to a transmission line or from a discontinuity, and cross-talk is the coupling of a signal on one line into an adjacent line.

14.2 Transmission Line Theory

14.2.1 PROPERTIES OF A SINGLE TRANSMISSION LINE

14.2.1.1. *Characteristic impedance*

A track running on one side of a printed circuit board which has a solid metal earth plane on its other side is called a microstrip line. Pulses can be sent along the line, down which they travel as a change in voltage and a change in current in the line. If the line is assumed to be lossless, it can be considered as comprising series inductance and shunt capacitance, both of which are uniformly distributed along its length. If the voltage impressed on the sending end of the line is fixed, the current in the line will be set by the characteristic impedance of the line, Z_0 such that $V/I = Z_0$. It can be demonstrated that in a lossless line $Z_0 = \sqrt{(L/C)}$, where L is the inductance per unit length and C is the capacitance. Since L and C both increase linearly with the length of the line, Z_0 is constant for any configuration of line, and is independent of the length of the line. For cases of practical interest to the designer of T.T.L. systems, Z_0 can be found from Fig. 14.1, or the formula $Z_0 = 11{\cdot}8L/\sqrt{\xi_\tau}$ can be used, where ξ_τ is the effective dielectric constant of the medium surrounding the line. This is not the same as the dielectric constant of the base laminate material from which the board is made, because only one side of the microstrip line is in contact with the base material; the other side and the edges are in air. On practical printed circuit boards, Z_0 lies between about 50 and 120 Ω.

14.2.1.2 *Propagation delay*

Pulses travel down the line, normally in the TEM mode, at a velocity which is a fraction (typically 0·6) of the velocity of the propagation of light. For lossless lines, $T_p = 1{\cdot}015\sqrt{\xi_\tau}$ nanoseconds per foot.

Values of T_p for lines likely to be used in the interconnection of T.T.L. gates can be found from Fig. 14.2. Typically, a value of 2 ns per foot of line can be taken for tracks on one sixteenth of an inch thick G10 epoxy-glass based board (see J. A. Scarlett *Printed Circuit Boards for Microelectronics* (Van Nostrand Reinhold, 1970), for details of board specifications).

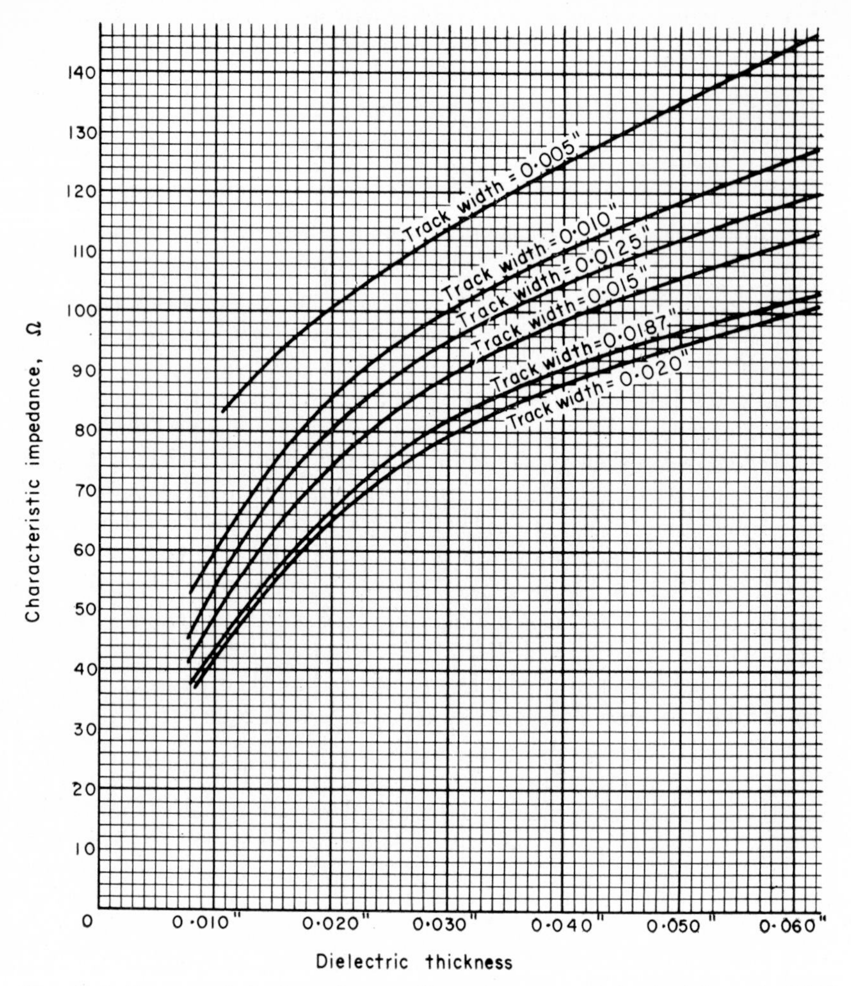

Fig. 14.1 Characteristic impedance of tracks on earth-backed boards (G10 epoxy-glass material).

14.2.2 REFLECTIONS IN A SINGLE TRANSMISSION LINE

14.2.2.1 *Terminations*

A pulse travelling along an infinitely long line 'sees' only the inductance and capacitance of the line ahead of it, and, in a lossless line, the propagation of the pulse along the line is uniform. However, on a finite line, the pulse will arrive at the receiving end of the line. What happens when it does so depends on the impedance it sees at the end of the line. If the line is terminated with an impedance Z_r equal to the characteristic impedance of the line Z_0, the change in current in Z_r associated with the change in voltage of the end of the line will be the same as the change in current along the line. In other words, the signal will be totally absorbed by the load.

If, however, the line is not terminated by an impedance equal to its characteristic impedance, the voltage and current at the end of the line

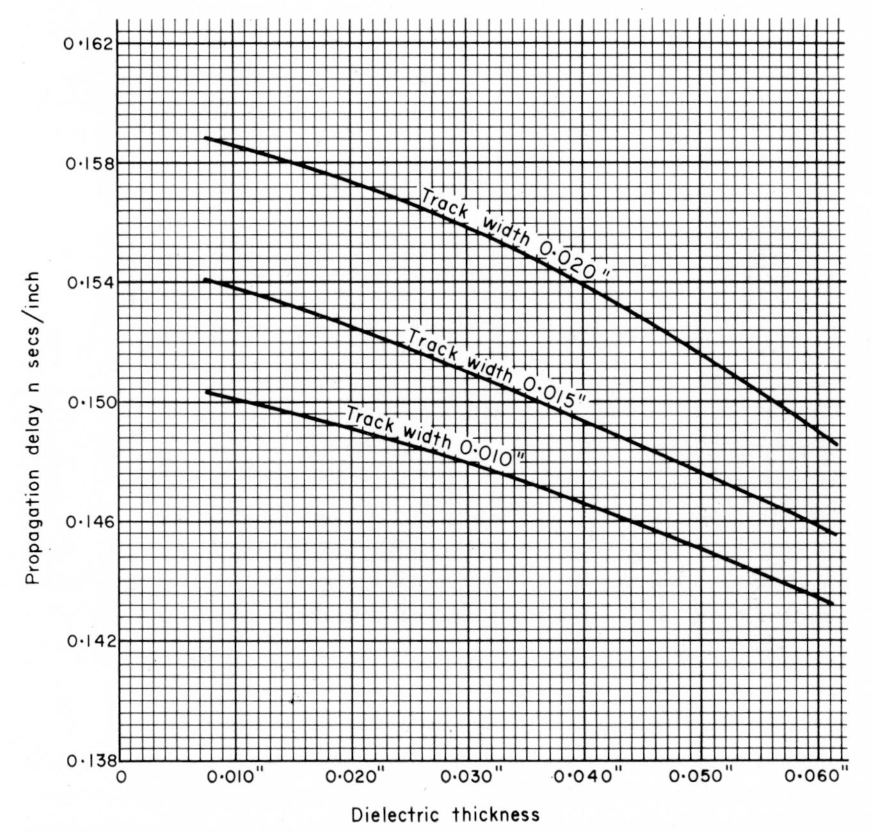

Fig. 14.2 Propagation delay on earth-backed boards (G10 epoxy-glass material).

cannot match those which have travelled along the line. The easiest cases to appreciate are the limit cases. If the end of the line is a short circuit, no voltage can exist at the end of the line, so the voltage pulse which has travelled along the line will be reflected back to the sending end in such a way as to leave the voltage at the receiving end of the line at zero. Alternatively, if the end of the line is an open circuit, no current can flow, so the reflection will be such as to leave zero current at the end of the line, and the voltage at the receiving end of the line will be doubled. For all practical cases, a partial reflection of the received signal occurs. When the terminating impedance is less than Z_0 the reflection reduces the voltage at the receiving end of the line and partially cancels the transmitted pulse as it (the reflection) travels back towards the sending end, whereas if the terminating impedance is greater than Z_0 the reflected wave increases the amplitude of the transmitted wave.

14.2.2.2 *Determination of Reflections*

Most transmission line circuits can be reduced to that shown in Fig. 14.3. In a circuit of this type, when the generator produces an ideal step of voltage E, at time t_0, a step of amplitude $V_{s1} = E[Z_0/(Z_0 + Z_s)]$ will be produced at point s. This step will travel down the line, associated with a

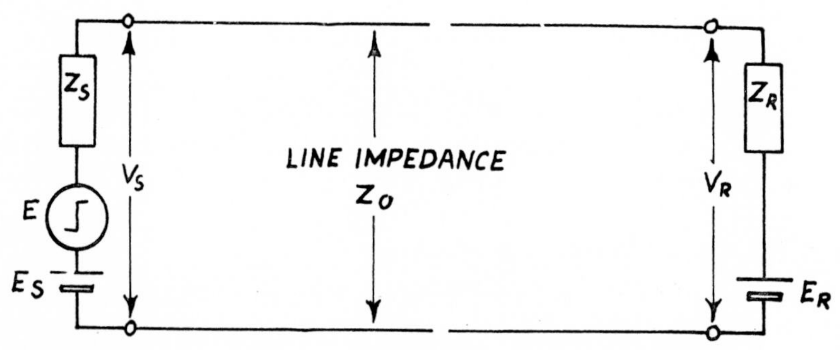

Fig. 14.3 Typical transmission line with terminations.

line current of V_{s1}/Z_0 and will arrive at point r after $T = 1T_p$ ns. Reflection will occur at r, and the reflection will be V_{r1}, ρ_τVsi where ρ_τ is the coefficient of reflection of the receiving end of the line

$$\rho_\tau = \frac{Z_r - Z_0}{Z_r + Z_0}$$

Values of ρ_τ vary from -1 when the load is a short circuit and the reflected wave cancels the incident wave, through 0 when the impedance matches the line impedance and there is no reflection, to $+1$ when the load is an open circuit and the line voltage is doubled on reflection.

The instantaneous change in voltage across Z_r will be

$$V_{s1} + V_{r1} \,(= V_{T1}) = V_{s1}(1 + \rho_\tau) = E\frac{Z_0}{Z_0 + Z_s}\left(1 + \frac{Z_r - Z_0}{Z_r + Z_0}\right)$$

The reflected wave travels back along the line to the sending end, where reflection again occurs, at time $2T$, or $2lT_\rho$ ns. The coefficient of reflection of the sending end is given by

$$\rho_s = \frac{Z_s - Z_0}{Z_s + Z_0}$$

The initial step at the sending end can be re-written as

$$V_{s1} = E\frac{(1 + \rho_s)}{2}$$

After reflection at the receiving end, $V_{s1}\rho_r$ is the wave which travels back to the sending end, and this results in the further reflection $V_{s1}\rho_r\rho_s$ which travels to the receiving end. The voltage at the sending end, V_s, changes at t_0, then at times $2T$, $4T$, $6T$, etc. as successive reflections occur, while the voltage at the receiving end, V_r, changes at times T, $3T$, $5T$, etc. The voltage at any point at any time is the sum of V_{s1} and all subsequent reflections which have occurred.

If the line is assumed to be lossless, V_s and V_r ultimately settle to a value of

$$V = \frac{EZ_r}{Z_r + Z_s}$$

The characteristic impedance of the line does not affect the ultimate voltage at its ends.

14.2.2.3 *Graphical solution of reflections*

The series of waves travelling along the line can also be found graphically, as shown in Fig. 14.4. Here, the line r1 represents the equation $V_r = E_r + Z_rI$; line r2 represents the equation $V_s = E_s - Z_sI$, and line r3 represents $V_s = E_s + E - Z_sI$.

Point A is then the steady state 'low' voltage, and point B is the steady state 'high' voltage. If the standing voltage generators E_s and E_r do not exist, the construction simplifies to that of Fig. 14.5, and the line voltage varies between 0 and E. Fig. 14.6 shows how the line can charge up from 0 to E volts. The generator first sees the line impedance Z_0, so a line is drawn on the graph from 0 to A, with a slope of $1/Z_0$. This line intersects the Z_s line at B, and this will be the amplitude of the initial step at the sending end of the line. From B, B–C is drawn with slope $-1/Z_0$. The intersection of B–C with 0–V at D gives V_{r1}. A further line from D at slope $1/Z_0$ crosses the Z_s line to give V_{s2}, from which point a further line at $-1/Z_0$ yields V_{r2}. This process is repeated until the steps become insignificantly small.

From this construction it can be seen that if Z_r equals Z_0 there will be no reflections. Figure 14.6 shows a situation where the line impedance is twice the impedance of the two terminations. This construction can be used for rising edges. For falling edges (falling to zero volts), the construction starts from V, and the sending end reflections occur on the line passing through the zero axes of the graph. Fig. 14.7 shows a construction for rising and falling edges in a case where Z_r is greater than the line impedance Z_0

but Z_s is less than Z_0. This figure also shows how the voltages at each end of the line can be plotted for the two cases to show the wave shapes at R and S.

This construction is valid if the terminating impedances are non-linear.

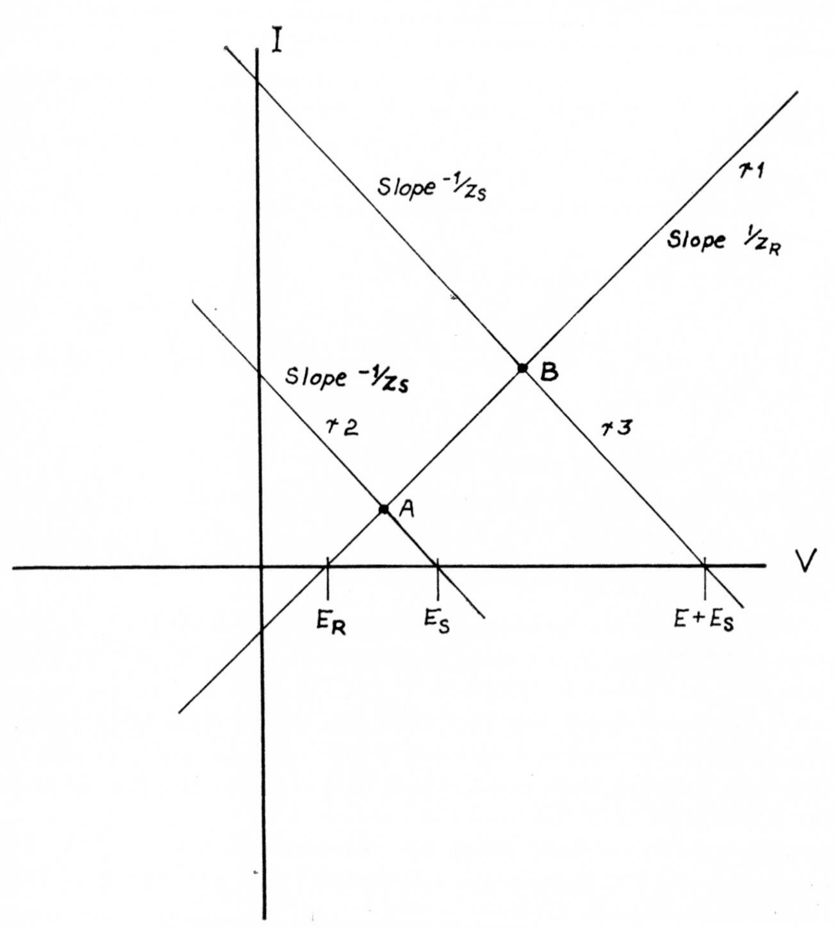

Fig. 14.4 Graphical solution for line voltages.

14.2.2.4 *Critical length*

So far, the edges of the pulses have been assumed to be virtually instantaneous. When the rise time of the pulse, t_R is much less than the delay down the line T, the only effect of a practical step with finite rise time is that the edges of all steps shown on a construction such as Fig. 14.7 will have a slight slope. The behaviour of the line is substantially unaffected by the edge speed until $t_R/T = 2$, or the rise time of the pulse is equal to

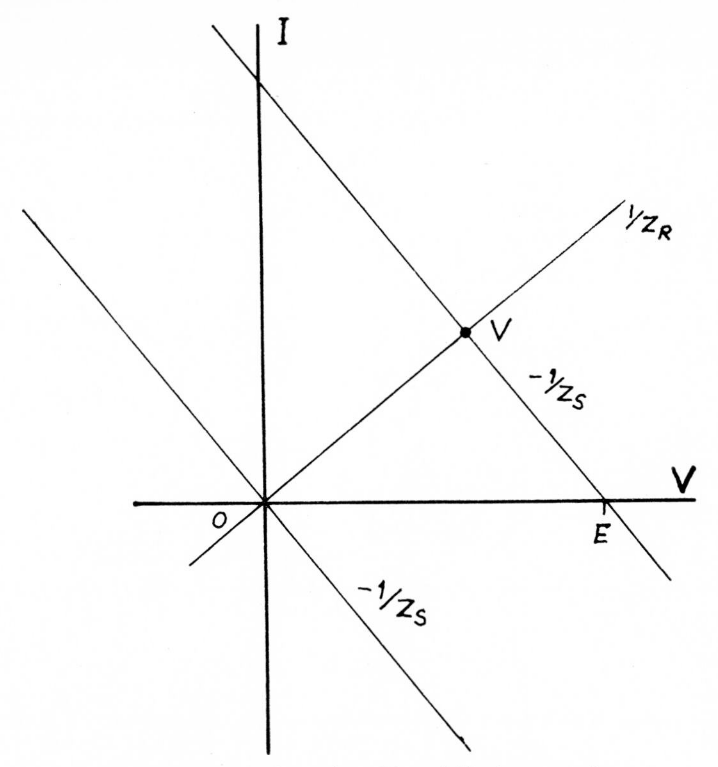

Fig. 14.5 Solution when E_s and E_r do not exist.

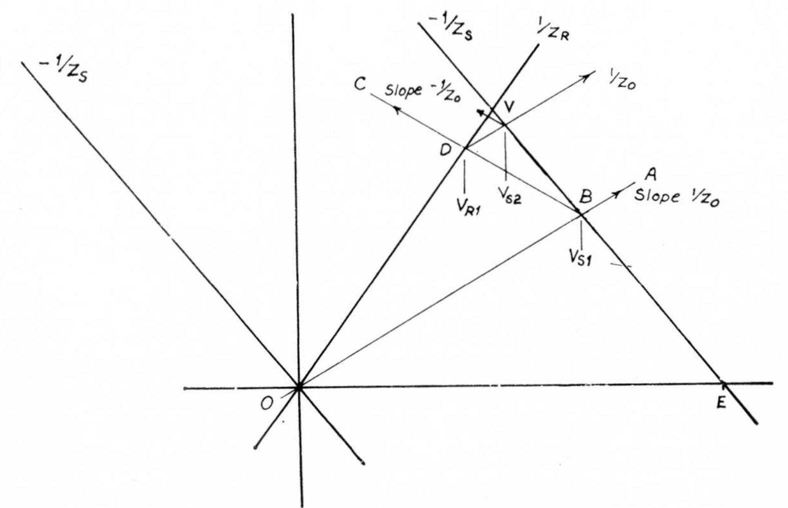

Fig. 14.6 Determination of reflections.

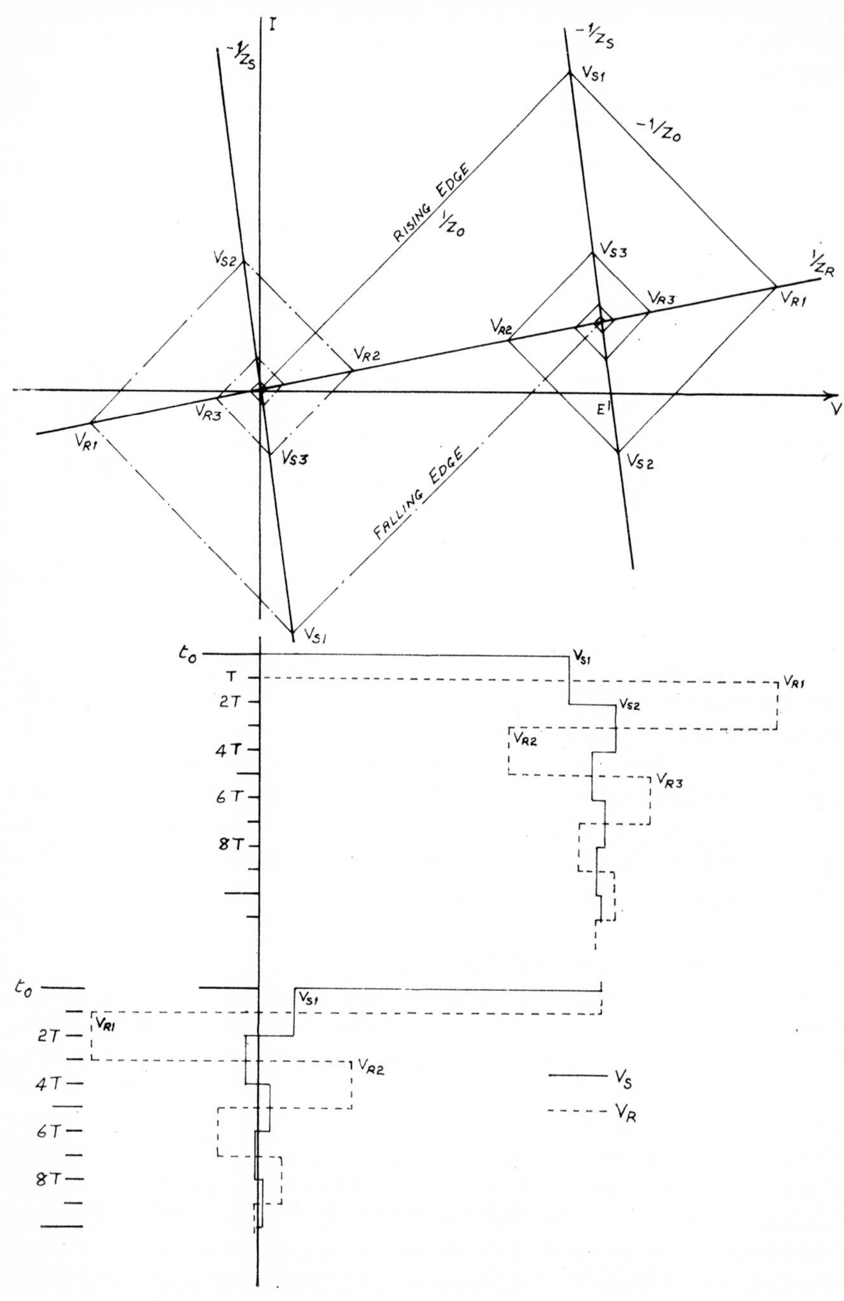

Fig. 14.7 Reflections on line where $Z_r > Z_0 > Z_s$.

twice the delay of the line. Then the first reflection will arrive back at the sending end just as the ramp reaches its final value for the step concerned. (See Fig. 14.8.)

As the line is made progressively shorter (i.e. as t_R/T increases), the reflections at each end will arrive before the final ramp amplitude has been reached, and the overswings and undershoots will be reduced in amplitude. Figure 14.8 shows cases where $t_R/T = 3$ and $t_R/T = 4$.

The length of the line is said to be critical when $t_R/T = 2$. that is

$$1_{cr}\,(\text{ft}) = \frac{t_R\,(\text{ns})}{2T_p\,(\text{ns/ft})}$$

Cases where t_R/T is less than 1 are called long lines, and cases where t_R/T is more than 4 are called short lines. It must be noted that a line can have an electrical 'length'—either critical, long, or short, only when the speed of the device which drives the line has been determined.

The response of a short line can be taken as being approximately the response that would be obtained if the line capacitance appeared as a lumped capacitance at the sending end.

14.2.3 DISCONTINUITIES

So far, the line has been assumed to be homogeneous; the inductance and capacitance per unit length have been assumed to be constant. This seldom occurs in practice. Any change in track width will form a discontinuity, as also will through-plated-holes, edge-connectors, branch lines or in fact any other change in the cross-section of the microstrip line.

At any discontinuity, there will be a reflection which will travel back towards the sending end, and the wave travelling towards the receiving end will also be modified by the discontinuity.

14.2.3.1 *Change in track width*

A change in the width of the track will mean that the characteristic impedance of the line will change at the discontinuity. The amplitude of the first reflection at this discontinuity can be found by taking the characteristic impedance of the line after the discontinuity as the load on the original section, and evaluating the reflection, and for the section after the change by taking the characteristic impedance of the main section of the line as the source impedance for the altered section of line. Such a simplification is effective only for the first reflection, because events at the discontinuity are complicated by the arrival of reflections back from the receiving end of the line. If there are several changes of track width along a line, it is simpler to treat the whole line as homogeneous with a modified characteristic impedance, unless the changes in width are widely separated

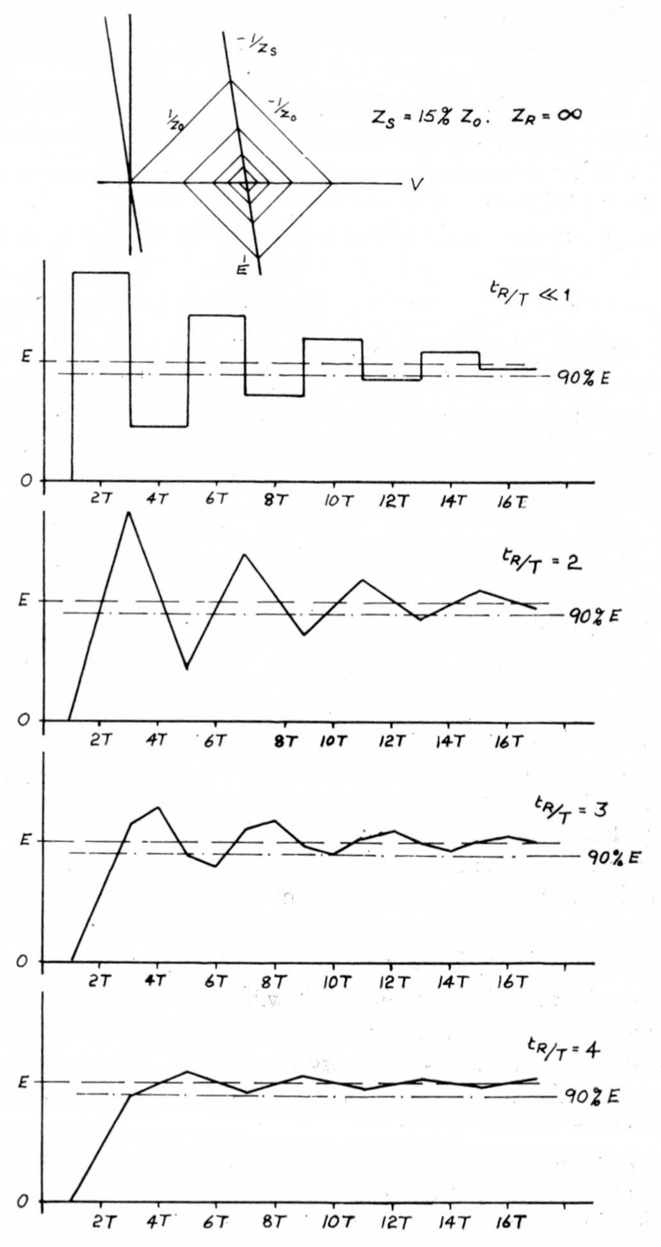

Fig. 14.8 Waveforms at receiving end of line for various values of t_R/T.

—i.e. unless the length of each section of the line is approaching the critical length for the system.

14.2.3.2 *Passive resistive discontinuities*

A similar approach to that adopted for changes in track width can be used for passive resistive discontinuities. Since the discontinuity will be seen in parallel with the line impedance, it must always appear to the wave as a resistive load of lower impedance than Z_0, so the effect will be to negate the signal in the line between the source and the discontinuity, and to delay the rise in the signal in the line beyond the discontinuity.

14.2.3.3 *Active resistive discontinuities*

A step of amplitude E which is propagating along a line will cause a change in current of E/R in an active resistive discontinuity of resistance R. This change in current will in turn result in a step in voltage on the line of

$$V = \frac{EZ_0}{2R}$$

because the current initially flows into the two parts of the line in parallel. This step of voltage V will propagate both ways along the line. Because the amplitude of the step produced is dependent only on the amplitude of the step on the line, the line impedance, and the resistance of the active discontinuity, it cannot be altered by altering the termination at either end of the line. These steps caused by active discontinuities can limit the permissible fan-out on a line.

14.2.3.4 *Capacitive discontinuities*

A capacitive discontinuity has an effect similar to that of an active resistive discontinuity since the capacity must be charged or discharged by current from the line. The effect is to negate the signal between the source and the discontinuity, and to delay the step after the discontinuity. Figure 14.9 shows the effect of a small capacitive discontinuity on a line of characteristic impedance of 100 Ω.

14.2.4 THE 'WIRED-OR'

A 'wired-OR' is a means of achieving a logical 'OR' function by wiring together the outputs of several devices (e.g. the collectors of transistors) such that when any device is ON, the whole node will be held in the ON state. A distributed wired-OR may present active resistive discontinuities to the line which interconnects the outputs. There is a much more serious effect to be considered when high speed logic circuits are being used. A line may be held low with more than one of the devices feeding the line

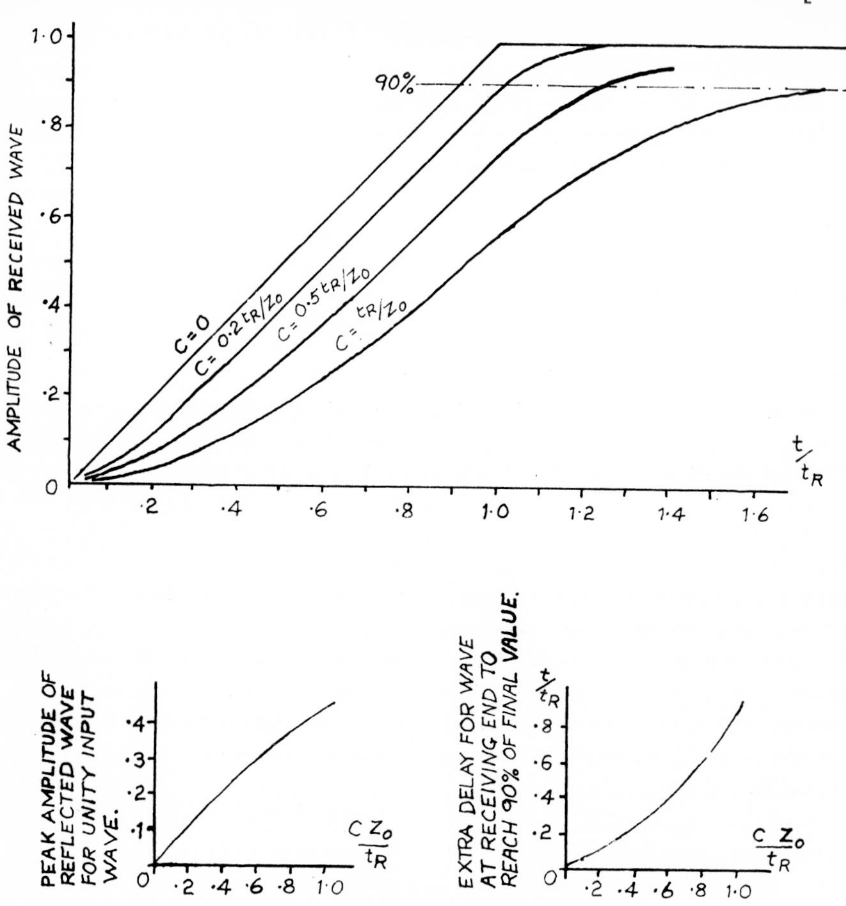

Fig. 14.9 Isolated capacitive discontinuity.

ON. If any one of these devices turns OFF, the node should remain low. However, it may be that all the current from the line is being sunk by one device, and this device is turned OFF. The current from the load being driven will then have to transfer, along the line, to another ON device. This redistribution of current among the outputs will result in voltage steps on the line, which may be detected by the inputs which the line feeds. The duration of these transients on the line will be a function of the separation of the outputs along the line, and for satisfactory working of fast logic circuits, this separation might have to be quite restricted. In particular, in the realization of highways where spatial distribution is a prime requisite, it might be possible to achieve only an 'exclusive-OR' function.

14.3 The T.T.L. Line

A T.T.L. line is a transmission line which interconnects T.T.L. gates. It exhibits the same properties as a plain transmission line, but the terminations at each end are non-linear.

14.3.1 REFLECTIONS

14.3.1.1 *Long line*

When an interconnection between two T.T.L. gates is of such a length that the full logic transition of the driving gate occurs before the first reflection returns from the receiving end, the interconnection is considered as a long line. The steps which occur at each end of such a long line could be calculated, but the non-linear nature of the terminations makes a graphical solution considerably simpler. A T.T.L. line is shown in Fig. 14.10, together with the full input characteristic of the driven gate and the high and low state output characteristics of the driving gate. (Leakage currents are assumed to be negligible—the input characteristic and the high level output characteristic continue to the right along the voltage axis.) Point A indicates the steady state 'low' condition, and point B the steady state 'high' condition. If a line impedance of 100 Ω is assumed, Fig. 14.11 can be drawn in a similar manner to Fig. 14.7. (See Section 14.2.2.3.) Points C and E indicate the level of the first two steps at the sending end of the line during turn-off, and D indicates the step at the receiving end. Because the high level output impedance of the driving gate used for this illustration matches the line impedance, there is only one step at the receiving end of the line. Points F and H indicate the first two steps seen at the driving end of the line during turn-on, and G and J indicate the steps at the receiving end.

In practice, considerable distortion will be seen on an oscilloscope trace. The stray capacitances of the set-up, together with the non-uniform rising and falling edges of the pulses from the driving gate will result in waveforms more like those shown in Fig. 14.12.

14.3.1.2 *Critical line*

As explained in Section 14.2, the length of an interconnection is said to be critical when the delay down the line is equal to half the edge speed of the driving gate, so that all voltage steps on the line have the same amplitude as on a long line, but there is no delay between the steps. Since it is seldom that a gate works with rising and falling edges of exactly the same speed, 'critical' lengths of line must cover a broad spread of lengths, over which some edges might cause full amplitude reflections but other

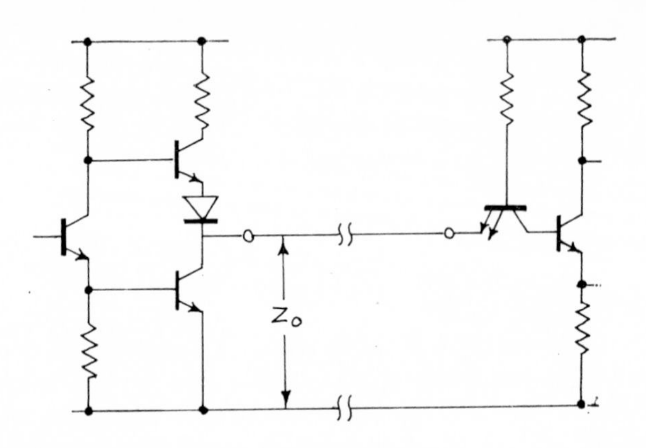

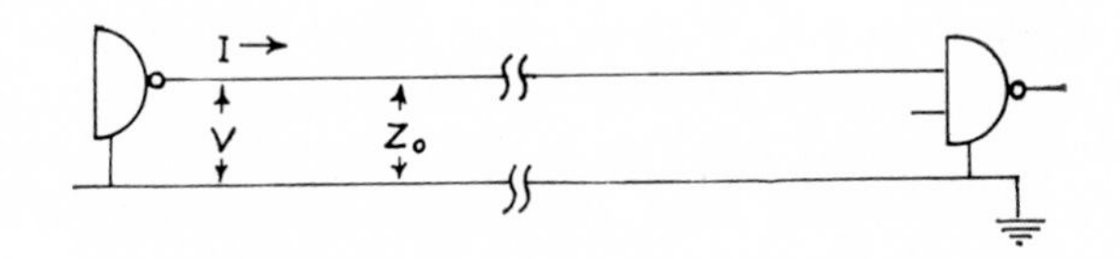

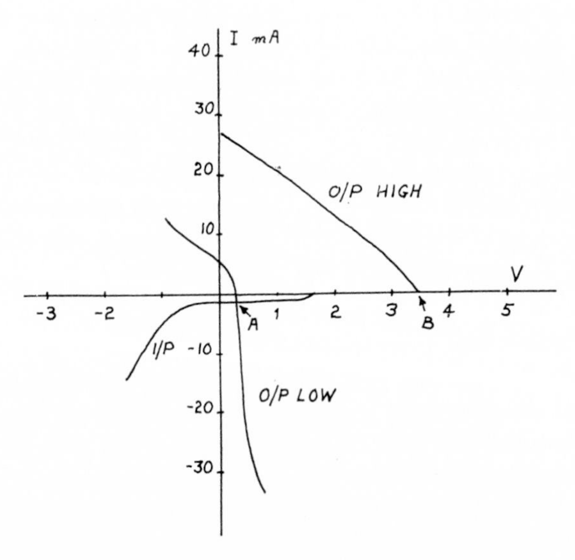

Fig.14.10 T.T.L. line and terminating impedances.

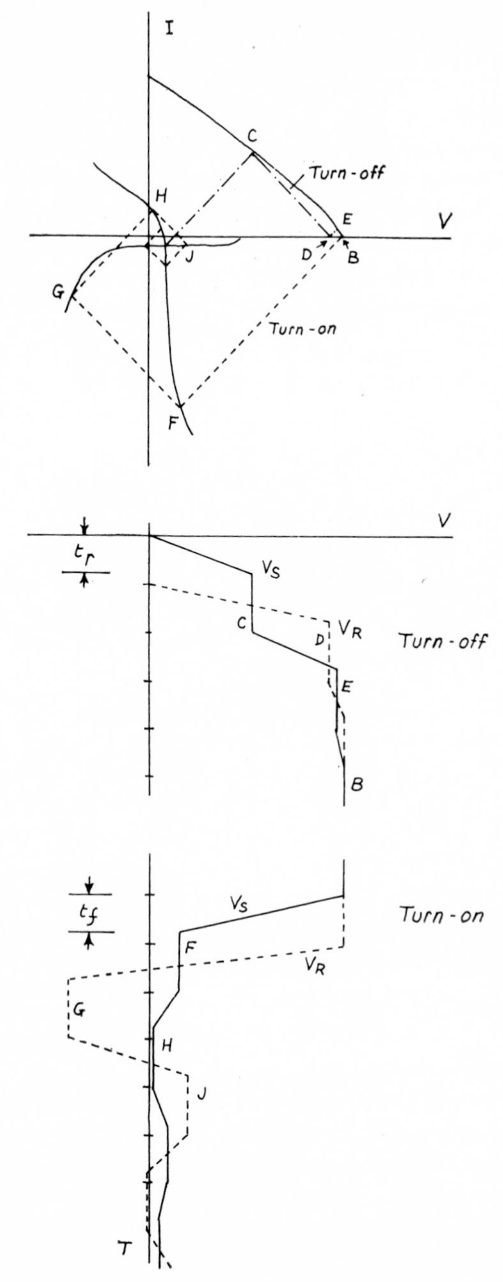

Fig. 14.11 Reflections on a T.T.L. line.

slower edges might not permit the steps to reach their full theoretical values.

14.3.1.3 *Short line*

As the length of the interconnection is decreased below the 'critical' length, the amplitude of the overswings and undershoots should decrease linearly to zero for a zero length interconnection. (See Fig. 14.8.) Since the critical length cannot be determined exactly, the amplitude of a reflection on a short line cannot readily be determined with any degree of precision; only an approximate estimate can be made.

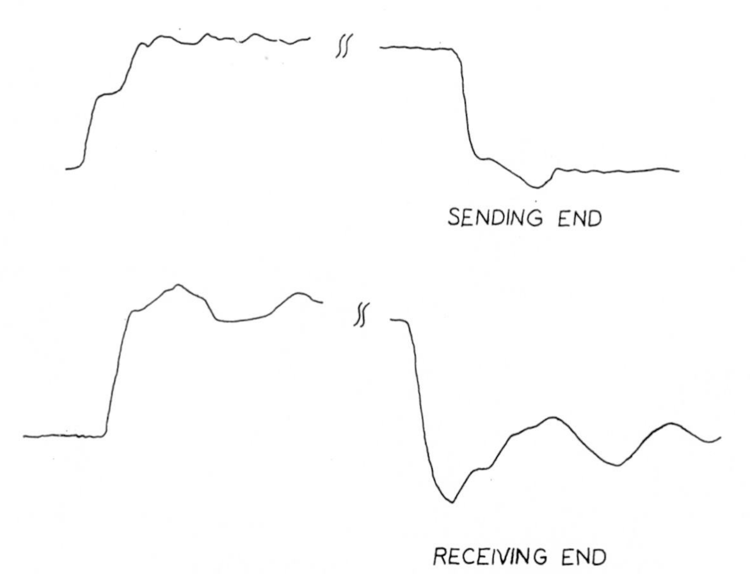

Fig. 14.12 Typical waveforms on a T.T.L. line.

14.3.2 AMPLITUDE OF STEPS

14.3.2.1 *Turn-off*

From Fig. 14.11 it can be seen that the amplitude of the first step at the sending end of the line during turn-off can lie near to the switching threshold of a gate input. This can mean that a gate fed from an adjacent gate which is also driving a long line cannot be guaranteed to switch until after the first reflection has come back along the line. That is, the apparent propagation delay will be increased by twice the line delay. The amplitude of this step depends on the line impedance, on the high level output impedance of the driving gate, and also on the fan-out into which it is driving. Fig. 14.13 shows how an increase in fan-out will raise the amplitude of this first step by increasing the current in the line and by also increasing the steady-state low voltage on the line.

For most practical cases, a worst-case output characteristic can be established from the specified high level output voltage V_{OH} and the minimum value of the output short circuit current I_{OS}. The higher the value of I_{OS}, the higher will be the amplitude of the first step.

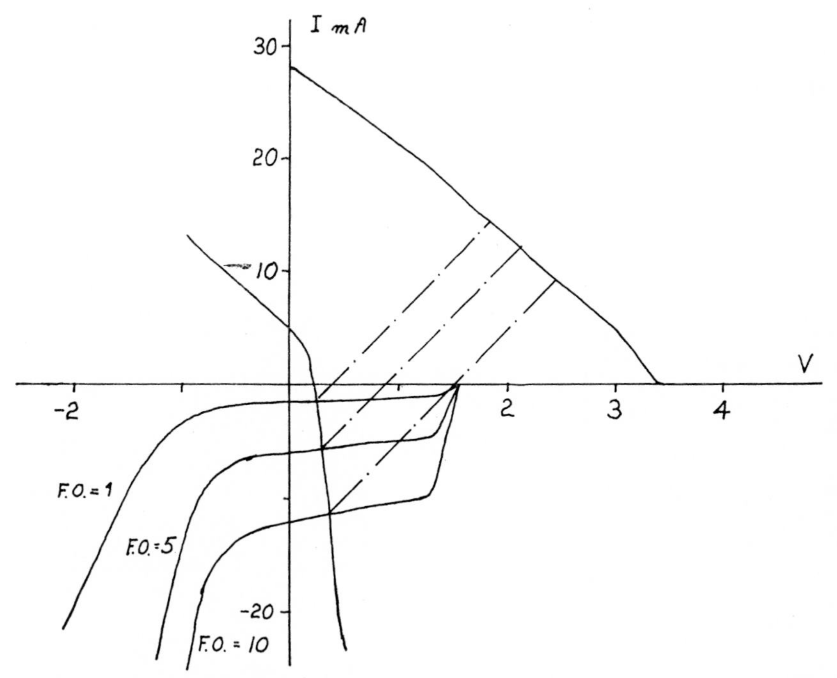

Fig. 14.13 Effect of fan-out on the first step at the sending end of the line during turn-off.

In the example used in Fig. 14.11, the output characteristic has a similar impedance to that of the line. Fig. 14.14 shows a case (a Series 2 gate) where the high level output impedance is considerably less than that of the line. The first step at the driving end of the line is still below the normal high level steady state voltage, but the first step at the receiving end of the line is well above the steady state voltage. The second step at the driving end is also high—with negligible leakage currents it will be virtually the same as the voltage at the receiving end. Successive reflections along the line slowly bring the line voltage down to the normal high level value. These steps cannot be drawn in at any reasonable scale, and the time required for the line to settle to its normal value is best determined from the lumped capacitance of the node and the values of the leakage currents.

This overshoot on a '1' level with a slow decay back to the normal value can lead engineers into suspecting that a fault has developed in an oscilloscope probe or in the switching of the oscilloscope amplifier—the effect is very similar to the appearance of a true square wave looked at with an a.c. coupled oscilloscope.

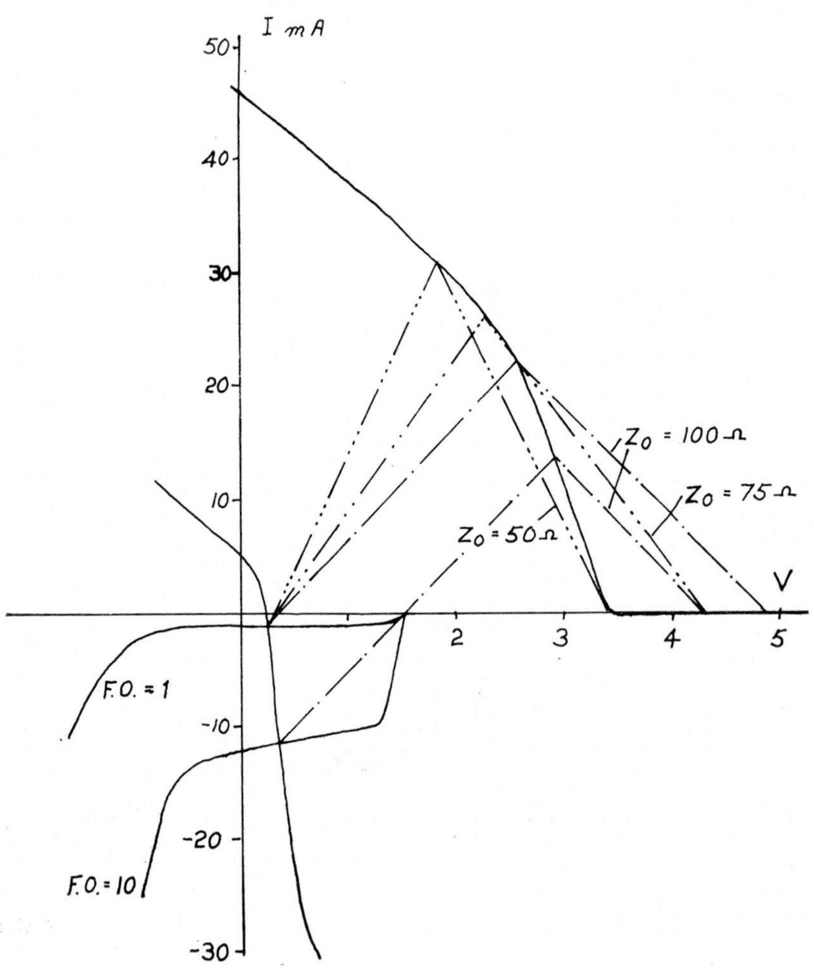

Fig. 14.14 Overshoot on turn-off—Series 2 gate.

These excessively high '1' levels are unlikely to damage a gate input, because if any input approaches breakdown the increased leakage current will discharge the line to a safe value. However, it is as well to avoid such high levels, as the turn-on time of the node will be increased, and cross-talk might further increase the line voltage to the point where the inputs have to handle an undesirably high current. These high '1' levels can be minimized by keeping the line impedance low (see Fig. 14.14). However, lowering the line impedance also reduces the amplitude of the first step at the sending end of the line. Increased fan-out raises the amplitude of the first step and also decreases the amplitude of the overshoot on the second step.

The first step at the receiving end of a line will always be near to (or above) the final '1' level, so correct switching can be guaranteed.

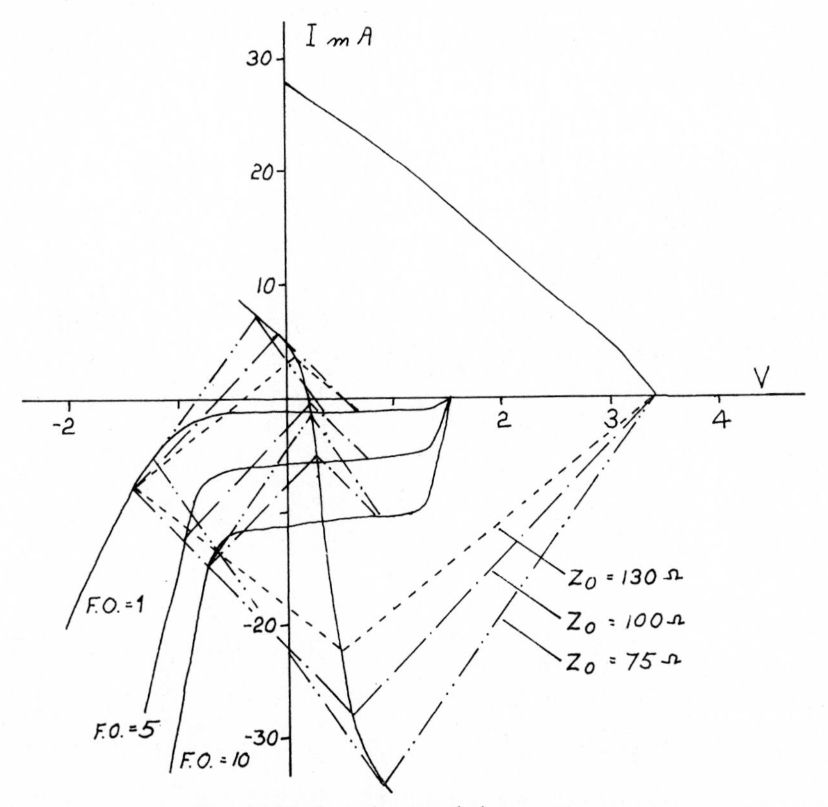

Fig. 14.15 Overshoot and ring on turn-on.

14.3.2.2 *Turn-on*

From Fig. 14.11 it can be seen that during turn-on the first step at both ends of the line is well below the normal input switching threshold level. The 'danger point' during turn-on is the first reflection at the receiving end (J), which is approaching the switching threshold level. Changes in fan-out or line impedance have little effect on the level of this reflection, as is shown in Fig. 14.15. The biggest effect of a change in fan-out is on the amplitude of the initial negative overshoot (G on Fig. 14.11).

The value of the negative overshoot is also affected by the impedance of the collector-substrate diode on the input transistor, or by the provision of input clamping diodes. Fig. 14.16 shows the values of overshoot achieved (with Z_0 of 100 Ω) when the input is clamped at about 1 V and when only a high impedance collector-substrate diode is available to limit the negative excursion. Although the values of the negative overshoot are quite

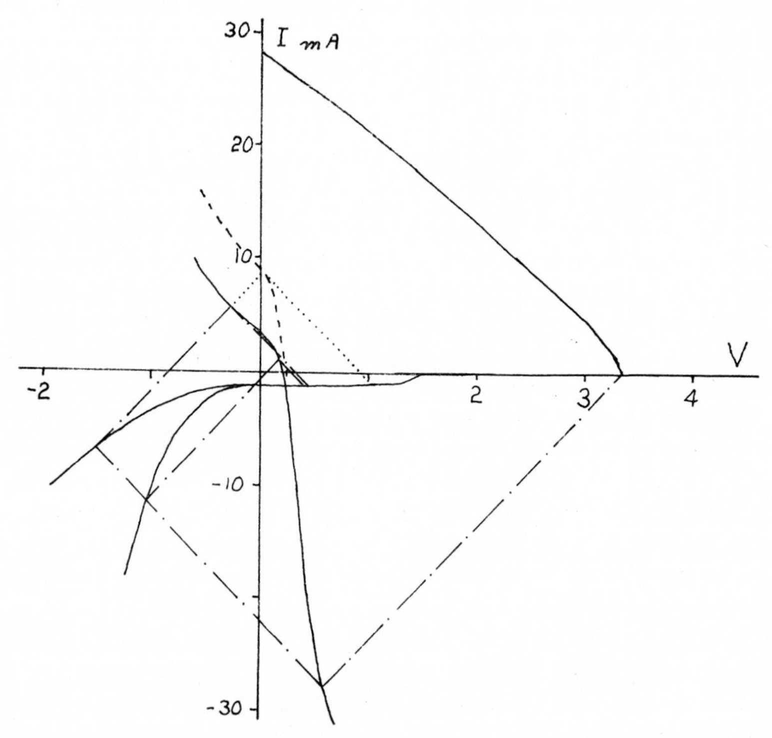

Fig. 14.16 Effects of input clamping and output characteristic on overshoot and ring during turn-on.

different, there is very little difference in the value of the positive going reflection (J) because of the reflection at the sending end of the line. It can be seen from Fig. 14.16 that the ability of the ON (low level) output to source current at positive voltages has a large influence on the value of the positive reflection at the receiving end of the line. A characteristic for a very poor output transistor is shown dashed, with the reflection from a 100 Ω line shown dotted.

It should be noted that throughout this section, the gate characteristics used for the illustrations are not actual measured curves; they have been drawn slightly distorted (especially in the negative voltage quadrants) to emphasize the reflections which normally occur.

14.4 Cross-talk

14.4.1 GENERAL CASE

Cross-talk is the noise induced into a line by the presence of a signal in an adjacent line. The simplest case is that of two 'long' parallel lines, both terminated in their characteristic impedance, with one line fed by an ideal generator with a pulse output of amplitude $2E$. The line which carries the signal from the generator is called the signal line, and the other line is the pick-up line. (See Fig. 14.17.)

14.4.1.1 *Back cross-talk*

As the rising edge (of amplitude E) occurs at the sending end of the signal line, (A), an edge of similar shape and of amplitude $K_B E$ is produced at the sending end of the pick-up line, (C). K_B, the back cross-talk constant, depends on the geometry of the system. For tracks between 0·010 in (0·25 mm) and 0·025 in (0·62 mm) wide, on normal printed circuit board material, K_B varies with board thickness and with the separation between the edges of the tracks as shown in Fig. 14.18.

A current of $K_B I = K_B E/Z_0$ flows in the pick-up line, in the reverse direction to the current I in the signal line. When this induced current wave reaches the receiving end of the pick-up line at D, a reflection occurs and a cancellation wave of amplitude $-K_B E$ travels back to the sending end of the line. After twice the time delay of the line ($2T$), this cancellation wave restores the voltage at C to zero.

If the signal pulse has a rise time of t_{TLH}, the total length of the pulse at C is $2T + t_{TLH}$. There is no disturbance at point D, and at points between C and D a pulse of amplitude $K_B E$ can be seen, the length of the pulse varying between zero and $2T + t_{TLH}$ as the distance from D varies.

As the length of the line is reduced, the amplitude of the induced pulse remains constant at $K_B E$ but the length of the pulse decreases until at the critical length the pulse at C is of triangular form, with amplitude $K_B E$ and duration $2t_{TLH}$. At intermediate points on a pick-up line of critical length both the amplitude and duration of the pulse are reduced linearly along the line.

On lines shorter than the critical length, back cross-talk will appear at point C as a triangular pulse, the amplitude of which will be below $K_B E$ as the length of the line is less than L_{cr}.

It should be noted that the induced pulse on the pick-up line will have some effect on the signal line. This effect is very small, and can safely be ignored for normal T.T.L. interconnections.

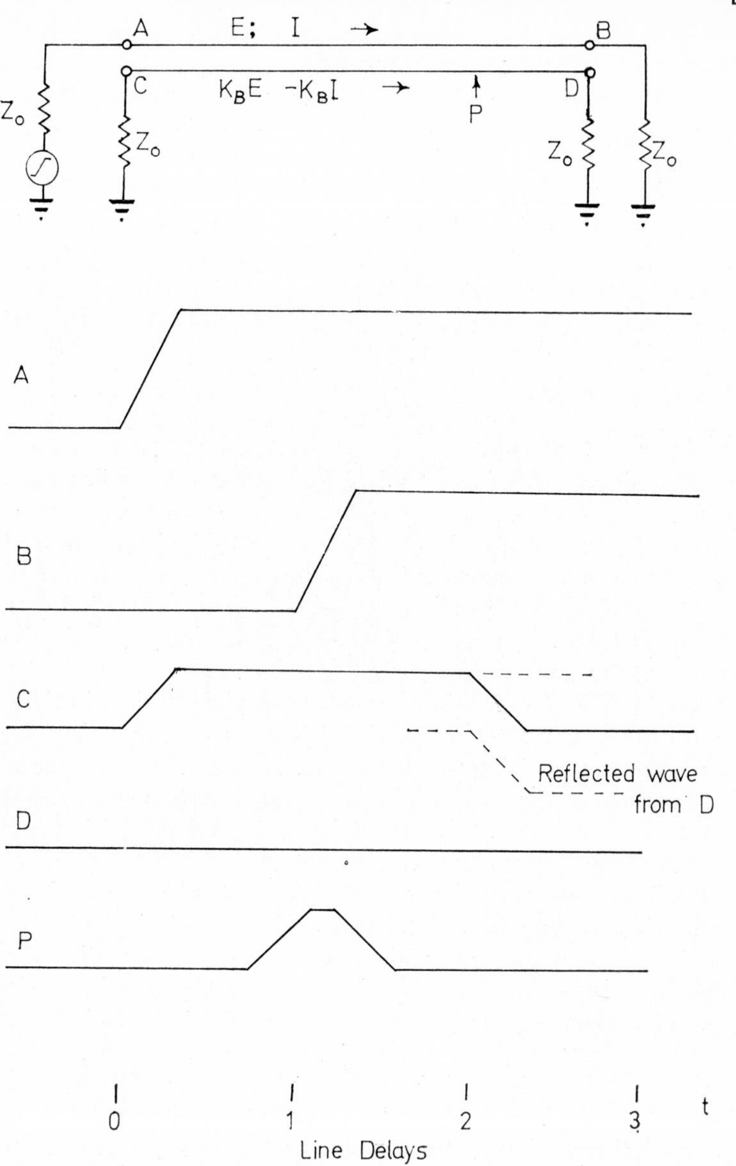

Fig. 14.17 Back cross-talk.

14.4.1.2 *Forward cross-talk*

In lines completely embedded in a dielectric material, this back cross-talk will be the only disturbance experienced by the pick-up line. However, tracks on the surface of a board can support two modes of propagation because the dielectric is partially the epoxy/glass base material and

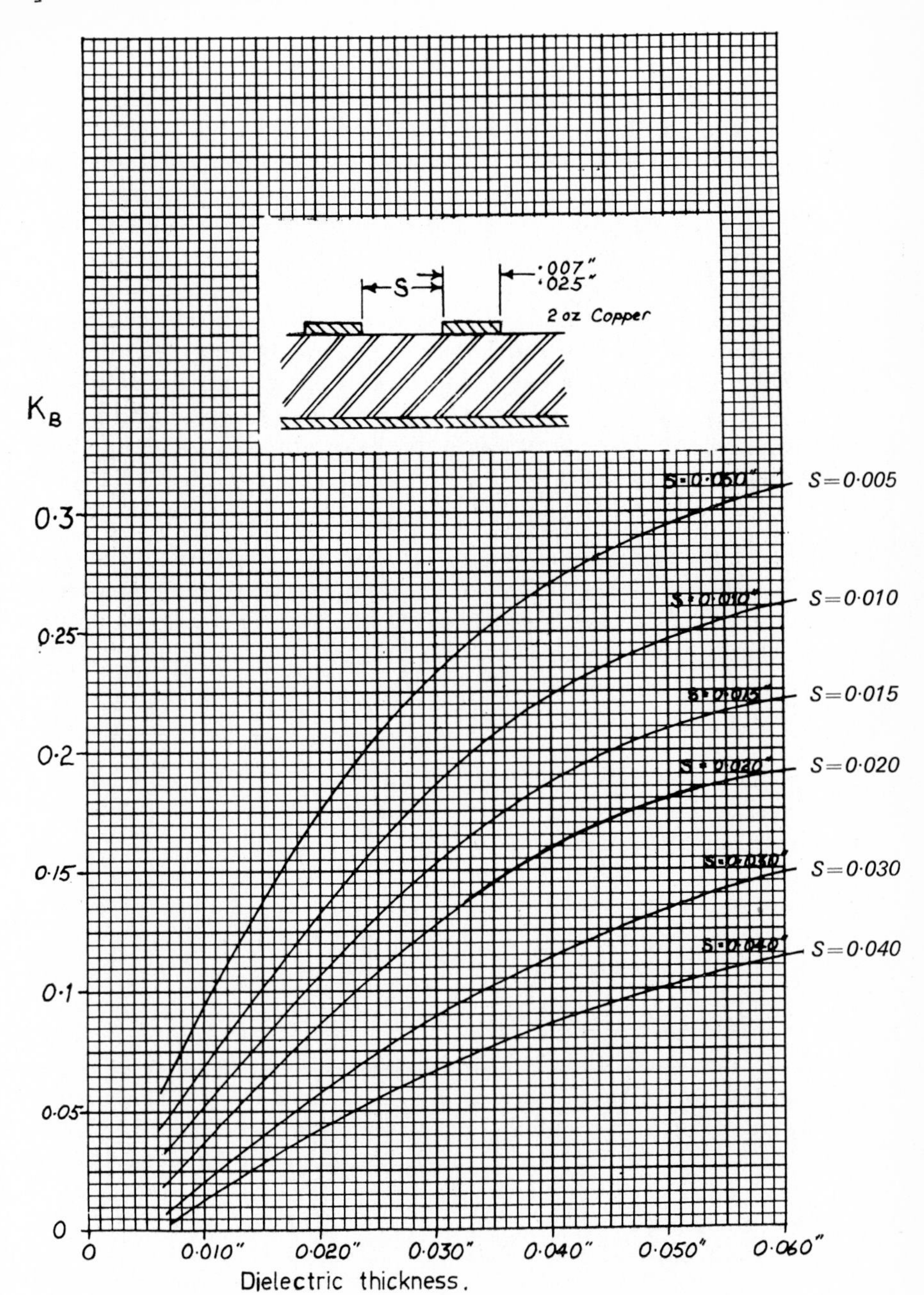

Fig. 14.18 Back cross-talk constant on G10 base boards.

partially air. Differential mode propagation between the two lines travels along the line faster than the common mode propagation between the line and the earth plane because the effective dielectric constant between the lines is lower than that through the board material. Thus a pulse sent along a microscrip line will have two components which arrive at the receiving end at different times, the difference in times being proportional to the product of the length of the line and the difference between the propagation delays.

If a rising edge of amplitude E is sent along the signal line, common mode propagation will result in a rising edge of amplitude $E/2$ on the pick-up line, whereas differential mode propagation will result in a cancelling falling edge of amplitude $-E/2$. Since the differential mode propagation is faster than the common mode, the result is a negative going pulse whose amplitude increases as the line length increases. The amplitude of this pulse can be found by

$$V_F = K_F l \mathrm{d}E_t/\mathrm{d}t$$

where l is the length of the line in feet, $\mathrm{d}E_t/\mathrm{d}t$ is the slope of the pulse edge in volts per nanosecond, and K_F is the forward cross-talk constant which can be found from Fig. 14.19. For most practical cases on printed circuit boards carrying T.T.L. devices, forward cross-talk can be ignored.

14.4.2 EFFECT OF PICK-UP LINE TERMINATIONS ON BACK CROSS-TALK

In Section 14.4.1 it was assumed that both lines were terminated in their characteristic impedance Z_0. If the sending end of the pick-up line is terminated in Z_0, and the line is long, the pick-up will be $K_B E$ at the sending end as described in Section 14.4.1.1 for any impedance at the receiving end. If the sending end termination Z_s has an impedance lower than Z_0, the amplitude of the pick-up pulse will be less than $K_B E$, whereas if Z_S is greater than Z_O, V_B will be greater than $K_B E$ until if Z_s is open circuit, $V_B = 2K_B E$.

If the impedance at the receiving end of the pick-up line, Z_r, is equal to the line impedance, only a single reflection will occur, but if Z_r is also mismatched, the pick-up line will ring in the same manner as a directly driven line. These reflections can be determined graphically as described in Section 14.2.2.3. In Fig. 14.20 a solution is shown for a case where $Z_s > Z_0 > Z_r$. The point of origin A is at $V = K_B E$ and $I = -K_B E/Z_0$. A negative going edge on the signal line would produce a similar set of reflections starting from the point $V = -K_B E$ and $I = K_B E/Z_0$.

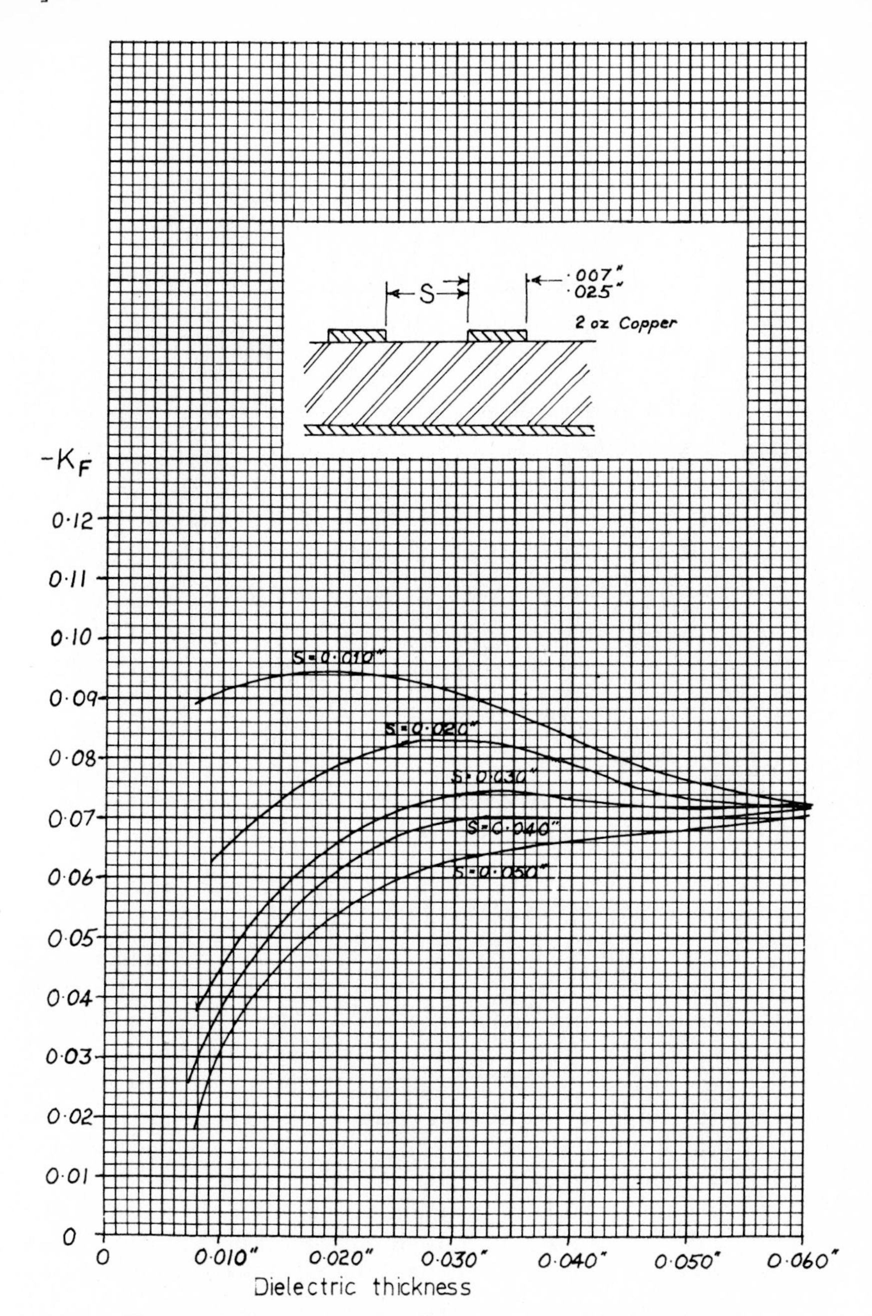

Fig. 14.19 Forward cross-talk constant on G10 base boards.

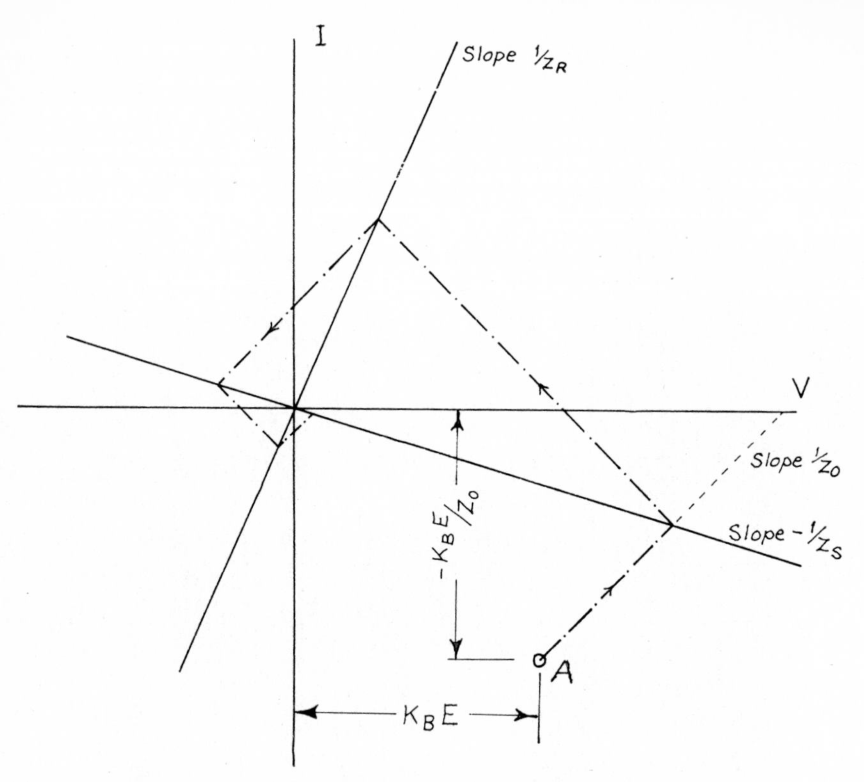

Fig. 14.20 Back cross-talk. Reflections on a mismatched pick-up line from a rising edge.

14.4.3 CROSS-TALK BETWEEN T.T.L. LINES

Cross-talk between T.T.L. lines can be calculated in the same manner as ring on a single line, but there are four separate steady-state conditions to be considered. The pick-up line may have its 'driving' gate at the same end as the signal line ('phase' condition), or the driving gates may be at opposite ends of the two lines ('antiphase' condition), as shown in Fig. 14.21. Also the output of the 'driving' gate on the pick-up line may be in the 'high' state or the 'low' state. The four steady-state conditions must be evaluated for rising and falling edges.

These cases can all be solved graphically, but when drawing out such solutions, care must be taken to work in the right quadrant of the graph, and to remember to add the induced voltage and current to the standing voltage and current already present on the line. Some engineers re-orientate the T.T.L. input and output characteristic correctly for each case to be solved. The author's preference is to work out all reflections with the characteristics drawn as in Fig. 14.10, merely reversing the current axis for the antiphase condition. Limit case characteristics for the logic family concerned can be drawn accurately on graph paper, and reflections for various values of K_B and Z_0 can be sketched on sheets of tracing paper laid over the characteristics.

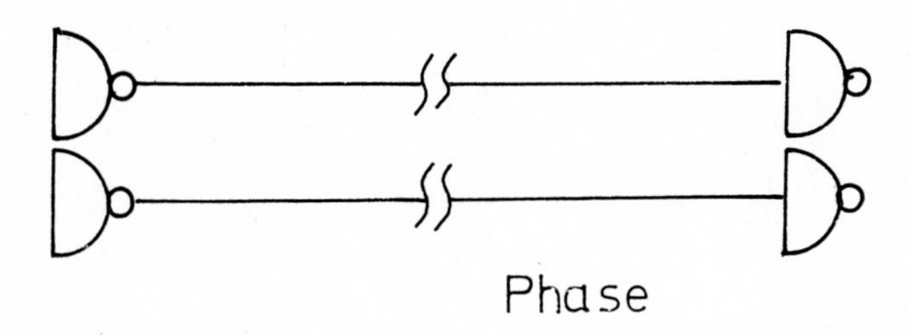

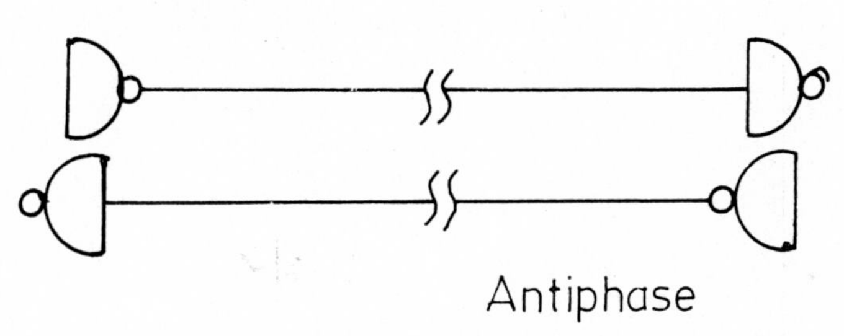

Fig. 14.21 T.T.L. cross-talk conditions.

14.4.3.1 *Pick-up line low conditions*

The reflections which occur on a long pick-up line which is at a logic '0' level are shown in Fig. 14.22 for phase and antiphase conditions, with both rising and falling edges. It can be seen that the antiphase condition with cross-talk from a rising edge is the only one of the four possible conditions in which the voltage on the input of the driven gate is likely to approach or pass through the switching threshold voltage. It must also be noted that point A is at $V_{OL} + K_B E$, and not at $K_B E$. Since the input impedance of the driven gate is high compared with Z_0, the amplitude of the first pulse at the sending end of the pick-up line will be approximately $2K_B E + V_{OL}$.

If the line impedance has been well chosen, it is most unlikely that either of the falling edge cases will cause the voltage on the gate input to approach the switching threshold. Similarly, the rising edge, phase condition is likely to be 'safe'.

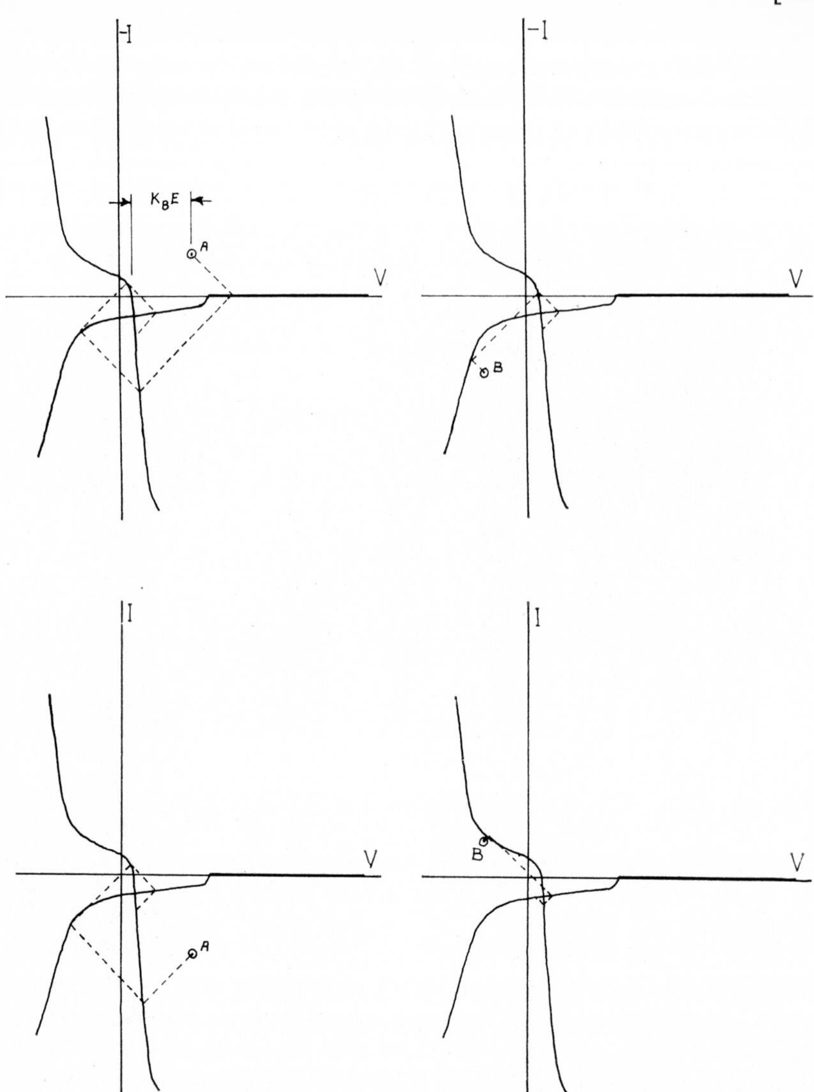

Fig. 14.22 Cross-talk with pick-up line low. (a) Rising edge, antiphase condition (b) Falling edge, antiphase condition (c) Rising edge, phase condition (d) Falling edge, phase condition.

14.4.3.2 *Pick-up line high conditions*

The reflections which occur when the pick-up line is in the logical '1' state are shown in Fig. 14.23. For both phase and antiphase conditions cross-talk from a rising edge will raise the voltage on the pick-up line to $2K_BE$ above V_{OH}, and the line voltage will decay slowly back to V_{OH} as described in Section 14.3.2.1 for ring during turn-off. In the antiphase condition, high input leakage current or even the start of input breakdown may limit the overall positive excursion on a rising edge, and in the phase condition, output leakage current may similarly limit the maximum positive voltage.

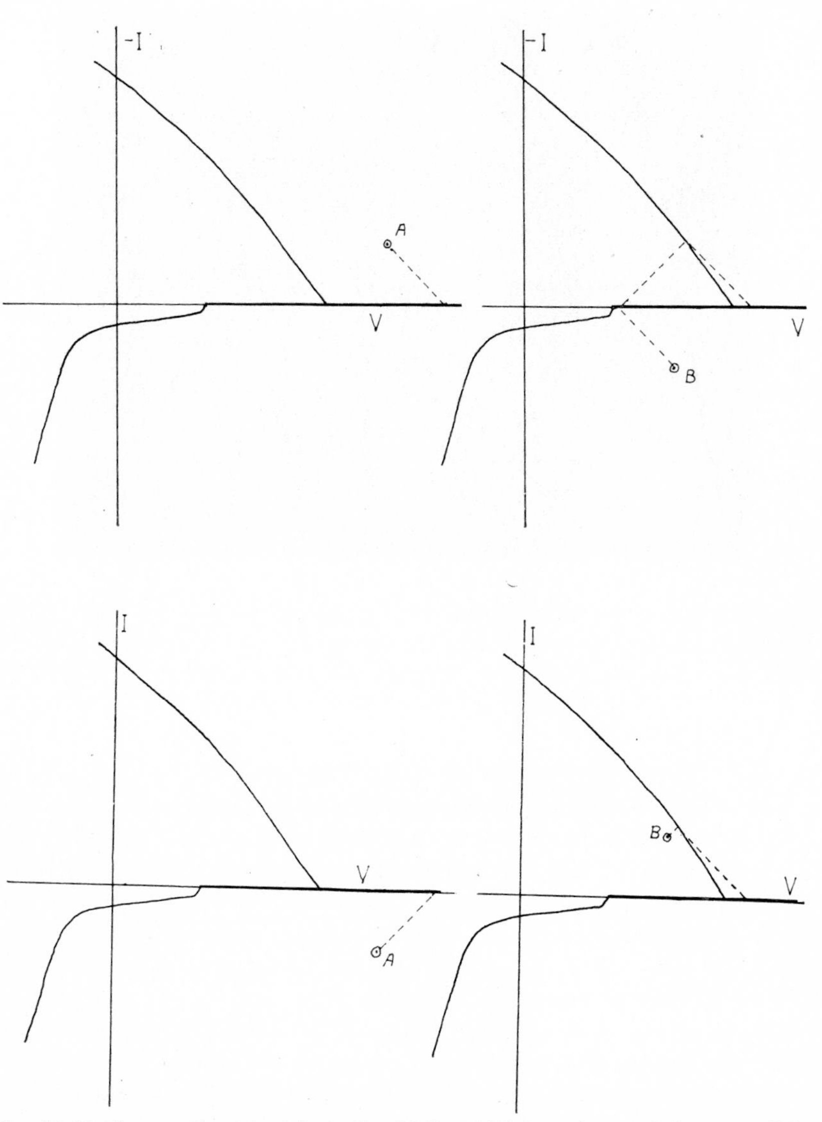

Fig. 14.23 Cross-talk with pick-up line high. (a) Rising edge, antiphase condition (b) Falling edge, antiphase condition (c) Rising edge, phase condition (d) Falling edge, phase condition.

Cross-talk from a falling edge in the antiphase condition can result in the voltage on the gate input approaching dangerously close to the switching threshold as shown in Fig. 14.23 (b). In the phase condition, the disturbance at the gate input (receiving end of the line) from a falling edge is small, but a negative pulse will be seen at the driving end of the line.

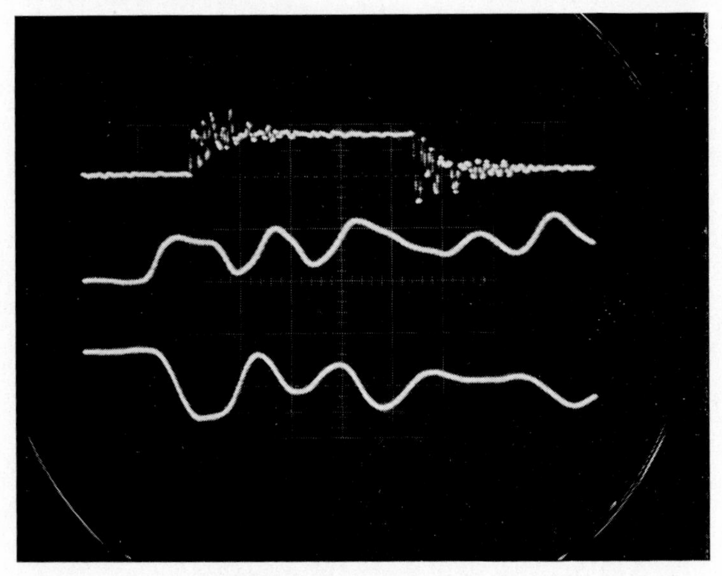

Fig. 14.24 Back cross-talk on high-level T.T.L. line 42 in long, Series 2 logic. All waveforms are 1 volt per cm. Top waveform is at 100 ns per cm. Two lower waveforms are expansions of rising and falling edges at 10 ns per cm. (Courtesy of the Marconi Co. Ltd.)

14.4.3.3. *Pick-up lines shorter than the critical length*

The reflections found from Figs. 14.22 and 14.23 apply when the lines are longer than the critical length. The duration of the reflections depends on the length of the line. For lines of less than the critical length, the pulses are triangular, and the amplitude of the pulse varies proportionately with the line length.

14.4.3.3 *Effects of ring on the signal line*

The foregoing description of cross-talk assumed throughout that the signal line experienced a single uniform rising (or falling) edge of amplitude E. In all practical cases, the signal line will ring as described in Section 14.3, and the resulting signal on the pick-up line will be the sum of the cross-talk signals induced by each successive reflected pulse on the signal line.

15

Cross-talk and Ring on Practical Printed Circuit Boards

15.1 Divergencies from Microstrip Lines

Chapter 14 explains the theory of ringing on an isolated microstrip transmission line, and cross-talk between a pair of lines. In any practical system, the cross-talk effects would be modified by the presence of other tracks, and calculations of ringing would be complicated by discontinuities (in the electrical impedance—not in the tracks themselves) at through-plated holes, board connectors, branching tracks, etc. Also, it might be economically desirable to use double-sided boards without any earth plane. A track on one side of a board with no earth plane on the other side to form a current return path is hard to visualize as a transmission line!

The difference in cost between double-sided boards and boards with internal earth planes, together with the difficulty in implementing any complex interconnection system on a board with only one side available for signal tracks, makes it very important to know whether microstrip line theory can be applied to double-sided boards. Two quite different kinds of boards must be considered: those which carry the T.T.L. packages, on which it is most likely that the tracks will be laid on *X–Y* co-ordinate principles; and back, or frame, wiring panels, on which it is most likely that the tracks on both sides of the board will run in the same direction. (For a full description of these and other types of board layout, see J. A. Scarlett in *Printed Circuit Boards for Microelectronics*, Van Nostrand Reinhold, 1970.)

15.2 Characteristic Impedance and Reflections on Double-sided Boards

15.2.1 EXPERIMENTS ON *X–Y* CO-ORDINATE BOARDS

Theoretical approaches to the question of the characteristic impedances of the tracks on double-sided boards became so complex that it was decided that a direct experimental approach would have to be used. Boards with tracks nominally 0·019 in wide on 0·050 in centres, and with pads 0·050 in diameter, had been made for an experimental T.T.L. system. (See Fig. 15.1.) Some of these boards were partially assembled with T.T.L. devices in such a pattern as to leave unconnected tracks which could be linked end to end at the unused package mounting pads. The resultant lines crossed themselves several times, and sometimes doubled back on their routes. In most cases the total length of line was about 2 ft which

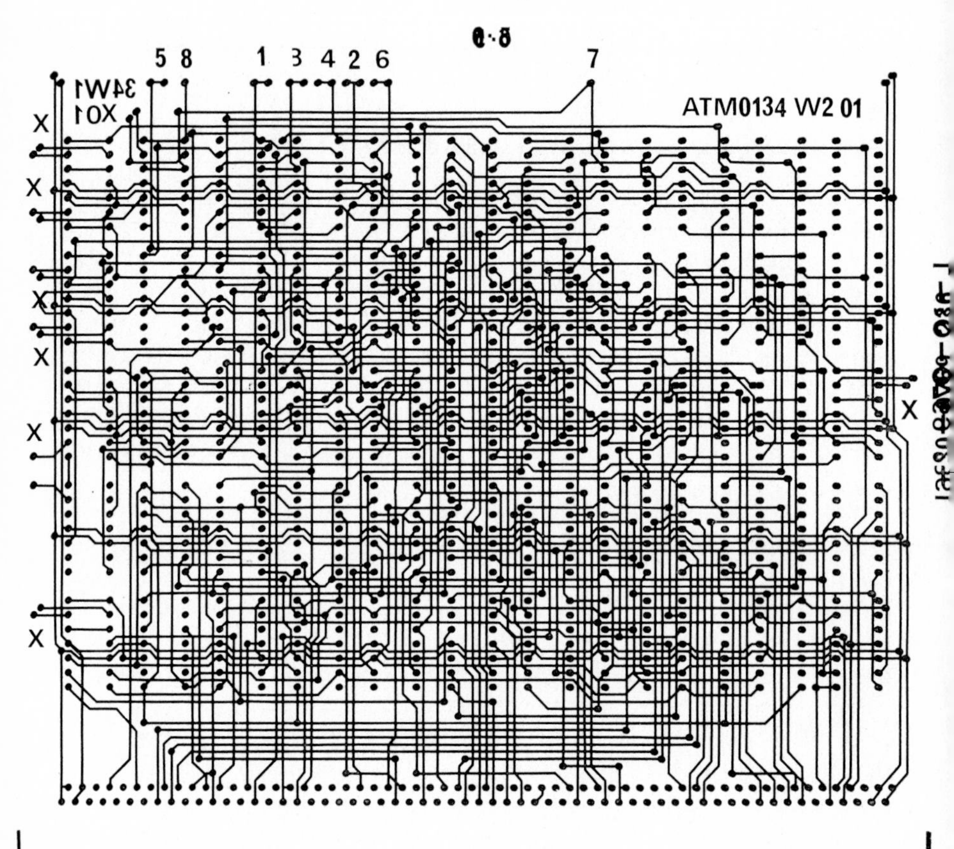

Fig. 15.1 Double-sided X–Y co-ordinate board (three-quarters of actual size). This illustration (which is printed from the photo-masters from which the boards were made) shows both sides of the board together, in registration. Tracks which run up and down the layout are on the upper or near side of the board, whilst tracks running across the layout are on the lower or far side of the board. The letters and figures which appear in this illustration as mirror images appear 'right reading' on the lower or far side of the board.

meant that by varying the edge speed of the output from a standard pulse generator, the lines could be tried in both 'long' and 'short' cases.

By varying resistive terminations to the lines until reflections were virtually eliminated (with power applied to the T.T.L. devices on the boards) it was found that all the lines tested appeared to exhibit a characteristic impedance of between 97 and 105 Ω. The uncertainty applied equally to all lines tested, and was considered to be caused by the numerous small discontinuities on the lines. Similar tests were tried with as much as possible of the logic on the boards switching, but although the results of these tests appeared to be similar, the virtual impossibility of totally shielding the oscilloscope probes and of eliminating all effects of cross-talk made the results less certain.

Further tests were tried with pairs of boards plugged into the back wiring panel, which also had similar sized tracks and pads to those on the boards. The main purpose of this series of experiments was to see whether the board connector used would introduce unacceptable reflections, but the results also demonstrated that when tracks on both sides of the board ran parallel without any earth plane the behaviour of the tracks could be predicted by the microstrip line theory.

15.2.2 EXPERIMENTS ON BOARDS WITH PARALLEL TRACKS ON BOTH SIDES

The results of the experiments on the X–Y co-ordinate boards showed that ring could be predicted for package-mounting boards, but it was felt that they could not be regarded as conclusive when applied to the back wiring panels with all tracks running in the same direction. Sets of boards were made from the pattern shown in Fig. 15.2. This had thirty parallel tracks on 0·050 in centres, with elongated pads to simulate the effects of the randomly distributed pads on practical boards. Boards were made with a similar pattern of tracks on each side; with tracks on one side and plain copper on the other side; and also with a 'cross' pattern of tracks on the 'back' of the board (Fig. 15.3). The boards were joined end to end by re-flow soldering fine wires to the pads, to make up lengths of up to 42 in.

The first series of tests on a single track in the centre of the pattern confirmed a characteristic impedance of 103 Ω on the 'earth-backed boards', with closely similar figures for the sets of boards with tracks on the back. The 'backing' tracks were connected to the outputs of unswitched T.T.L. gates, alternate tracks fed from opposite ends of the boards, with alternate gates at each end 'low' (or 'high'). The 'receiving' ends of the backing tracks were left open circuit.

Tests were run with the signal generator driving the line under test, and also T.T.L. gates (both Series 1 and 2) whose output characteristics

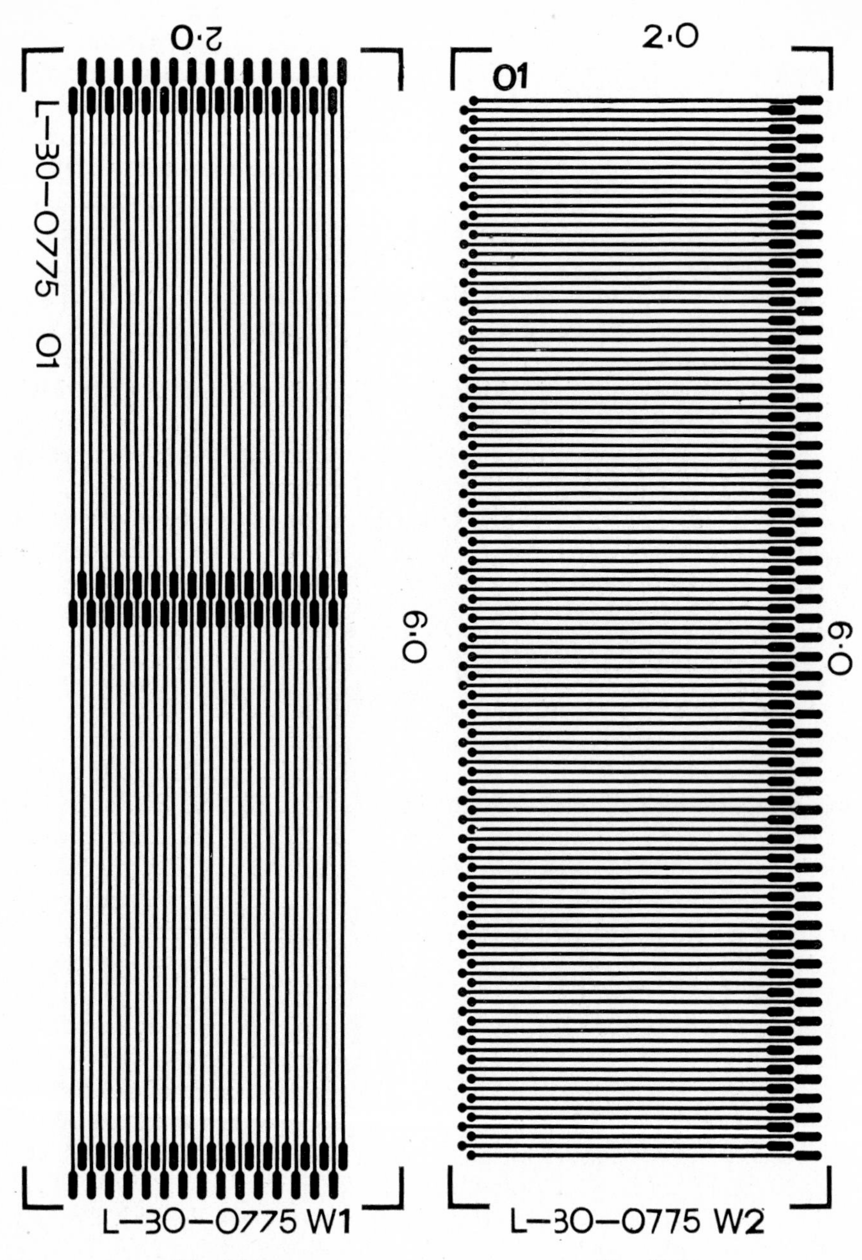

Fig. 15.2
Parallel-track test board.

Fig. 15.3
Cross-track backing for test board.

had been accurately measured. In all cases the reflections were as predicted by the theory.

It was also found that up to 75 per cent of the 'backing' tracks could be peeled off the boards without any significant effect on the reflections or on the characteristic impedance as measured by varying the value of the terminating resistors.

15.3 Cross-talk in Multiple Lines Configurations—Experiments

15.3.1 CO-RELATION BETWEEN EXPERIMENTAL RESULTS AND THEORY

The boards used for the characteristic impedance and reflection tests were also used to measure cross-talk in a variety of multi-track configurations.

An initial series of tests in which only two tracks were used confirmed that microstrip line theory could be applied to boards with parallel or cross-backing tracks connected as described in Section 15.2.2. These initial tests also confirmed that 10 Ω resistors could be used as adequate substitutes for ON gates to drive the 'backing' lines. The effects of variations in track widths and spacings were verified by tests on boards with groups of tracks 0·005, 0·010, and 0·015 in wide on 0·025 in centres, and 0·010, 0·015, and 0·025 in wide on 0·050 in centres (see Fig. 15.4).

15.3.2 NOMENCLATURE USED IN MULTIPLE LINE CONFIGURATIONS

When only two lines are used, they can be unambiguously described as the 'signal' and the 'pick-up' line, and the disposition of their terminations is defined by 'phase' and 'antiphase' conditions. As more lines are considered, descriptions of the condition being tested become quite unwieldy, and so a simplified notation was used to describe the various configurations tried. All multiple line configurations were tried only in the antiphase condition, so S and P describe completely the signal and pick-up lines. A line held in a 'high' state by a gate at the sending end (of the signal line) is defined by H, and L defines a line driven by a 'low' gate at the sending end. Lower case letters define lines held low or high by gates at the receiving end. Lines held low with a resistor at the sending end are defined by a, and b defines a resistor at the receiving end. E defines a line earthed at both ends.

By writing the appropriate letters in the order in which the tracks run, the most complex systems can be described. A single row of letters represents tracks on earth backed boards, or on boards with the backing tracks running across the signal tracks. On boards with tracks running parallel on both sides, two rows of letters are used. When a pattern of tracks

Fig. 15.4 Portion of test board with clearances from 0·010 in to 0·040 in.

is repeated out to the edges of the board, dashes are used. Thus the cases already described are:

$$\text{S P} \qquad \frac{\text{S P}}{\text{— H l L h H l L —}} \quad \text{and} \quad \frac{\text{S P}}{\text{— E E E E —}}$$

The results of all tests were normalized to the SP case so that for each configuration a constant, K_C, could be found by which K_B could be multiplied to predict cross-talk in practical designs.

15.4 Cross-talk in Multiple Lines

15.4.1 EFFECT OF NON-SWITCHING LINES ADJACENT TO SIGNAL AND PICK-UP LINES

The first multiple line cases to be considered are those in which the signal and pick-up lines are in the middle of a group of non-switching lines.

On earth-backed boards, a line in the logical 'high' state either side of the pick-up and signal lines has very little effect on the cross-talk, and the configuration H S P H can be regarded as having a K_C of 1·0. Lines in the 'low' state either side of the pair of lines under consideration reduce the pick-up considerably when the driving gates for the non-switching lines are at the 'sending' end of the system; K_C for the case L S P L is 0·6. Earthing the pair of tracks outside the pair of lines under consideration halves the cross-talk; K_C is 0·5 for E S P E.

On boards with parallel running tracks on both sides, the cross-talk between tracks on one face varies depending on whether the tracks on the other face are driven from the sending or receiving end of the board. When all the 'backing' tracks are held low at the sending end, the cross-talk is only slightly greater than when earth-backed boards are used, but when the 'backing' tracks (non-switching) are driven from the receiving end, the cross-talk increases by about 50 per cent.

Values of K_C for cases of one signal and one pick-up line in a group of parallel running tracks on both sides of the board vary from 1·0 for the case

$$\frac{\text{— b b S P b b —}}{\text{— b b b b b b —}} \quad \text{down to 0·6 for the case} \quad \frac{\text{— a a S P a a —}}{\text{— a a a a a a —}}.$$

15.4.2 EFFECT OF MULTIPLE SIGNAL TRACKS SWITCHED SIMULTANEOUSLY

As more signal tracks are run in an adjacent group about the pick-up line, the cross-talk increases, up to a maximum when about ten tracks are switching together, for which the value of K_C is about 3. It was considered that the measured results for groups of more than six signal tracks were of doubtful value, as probe pick-up and small differences in the speeds of the driving gates made the readings from successive experiments inconsistent. The figure 3 for a group of ten lines represents the highest value observed. Values for groups of up to six signal tracks are:

	K_C
S P S	1·45
S S P S	1·7
S S P S S	2·0
S S P S S S	2·2
S S S P S S S	2·4

15.4.3 EFFECT OF SIGNAL TRACKS ON BOTH SIDES OF THE BOARD

In cases where signal lines are run on both sides of the board, the cross-talk varies with the number of signal lines in the same way as when all the signal lines are all on the same side of the board as the pick-up line. This relationship will apply only when tracks about 0·020 in wide, running on 0·050 in centres are used on one sixteenth of an inch thick G10 based material. If the track spacing is varied, K_B for the tracks on the face of the board will vary as in Fig. 14.18, but the cross-talk between tracks on opposite sides of the board will vary only slightly. If the thickness of the board is reduced, K_B for tracks on one face of the board will be decreased, and the cross-talk will reduce, but coupling through the board will be greater.

For tracks 0·019 in wide, separated by 0·031 in, on 1/16 in G 10 material:

		K_C
— b S P b — / — b b b b —	— b b S b b — / — b b P b b —	1·0
— b S P S b — / — b b b b b —	— b S P b b — / — b b S b b —	1·5
— b S S P S S b — / — b b b b b b b —	— b S P S b — / — b S S b b —	2·0
— b S S S P S S S b — / — b b b b b b b b b —	— b S S P S b — / — b b S S S b —	2·5

15.4.4 EFFECT OF INTRODUCING EARTH LINES EITHER SIDE OF THE PICK-UP LINE

The effect of an earth line either side of the pick-up line is to halve the value of the cross-talk experienced in multiple signal-line configurations. Thus case S P S has $K_C = 1{\cdot}45$, and S E P E S has $K_C = 0{\cdot}7$.

Mixtures of earth and signal tracks on the back of the board similarly reduce the cross-talk. No significant difference could be seen whether the track directly opposite the pick-up line was carrying a signal or was earthed. Thus

$$\frac{\text{— b b S E P E S b b —}}{\text{— b b b S E b b b b —}} = \frac{\text{— b b S E P E S b b —}}{\text{— b b b E S b b b b —}} \quad K_C = 0{\cdot}85$$

15.4.5 EFFECT OF UNSWITCHED TRACKS BETWEEN SIGNAL AND PICK-UP LINES

The introduction of unswitched tracks between the signal and pick-up lines reduces the pick-up when the pick-up line is 'low', but has less effect when the pick-up line is 'high'.

Unswitched lines held 'low' at the sending end give very good screening. Lines held 'high' at the sending end or 'low' at the receiving end are less effective, and lines held 'high' at the receiving end have only a slight effect on the cross-talk.

	K_C
S E P E S	0·7
S L P L S	0·7
S l P l S	1·0
S H P H S	1·0
S h P h S	1·2
S P S	1·45

If two unswitched lines are run between the signal and pick-up lines, the screening effect is increased. If the two unswitched lines differ in phase or the state of the gates which drive them, the screening achieved is mainly dependent on the condition of the lines adjacent to the signal lines.

	K_C
$\dfrac{\text{— b S L h P h L S b —}}{\text{— b b b b b b b b b —}}$	0·7
$\dfrac{\text{— b S h L P L h S b —}}{\text{— b b b b b b b b b —}}$	0·9

Table 15.1 summarizes values of K_C for a number of track configurations likely to be met with in practical designs. In cases where the 'back' of the board is held 'low' at the receiving end, the cross-talk will be reduced if some of the 'backing' tracks are held low at the sending end. Cross-talk will increase if a significant number of the 'backing' tracks are in the 'high' state.

TABLE 15.1 Values of K_C for various track configurations (0·019 in wide tracks on 0·050 in centres on $\frac{1}{16}$ in G10 material)

K_C	Earth-backed boards	All backing lines held steady	Signal tracks both sides	Interposed logic tracks	Interposed earths
0·5	E S P E				
0·55				— b S L L P L L S b — — b b b b b b b b b —	
0·6	L S P L	— a S P a — — a a a a —			
0·7	S E P E S S L P L S			— b S L h P h L S b — — b b b b b b b b b — — b S L P L S b — — b b b b b b b —	— b S E P E S b — — b E S E S E b — — b S E P E S b — — b b b b b b b —
0·8					— b S E P E S b — — b E S E b b b —
0·85					— b S E P E S b — — b b S E b b b —
0·9				— b S h L P L h S b — — b b b b b b b b b —	

1·0	S 1 P 1 S H S P H S P S S E P E S S S H P H S	— b P b — — b S b — — b S P b — — b b b b —		— b S 1 P 1 S b — — b b b b b b — — b S H P H S b — — b b b b b b b —	— b S E P E S b — — b b b S b b b —
1·2	S h P h S			— b S h P h S b — — b b b b b b b —	
1·45	S P S				
1·5		— b S P S b — — b b b b b —	— b S P b — — b b S b —		
1·7	S S P S				
2·0	S S P S S	— b S S P S S b — — b b b b b b b —	— b S P S b — — b S S b b —		
2·2	S S P S S S		— b S P S b — — b S S S b —		
2·4	S S S P S S S				
2·5		— b S S S P S S S b — — b b b b b b b b b —	— b S S P S b — — b b S S S b —		

15.5 General Notes on Cross-talk on Double-sided Boards

It is a fairly obvious point that the introduction of earth tracks into a multiple lines configuration will reduce the cross-talk, and probably equally obvious that in equipments of any size space is unlikely to be available for many such earth tracks. Unswitched tracks in the 'low' state give fairly good screening, but it must be remembered that a track which is unswitched at one time under consideration will switch at some other time, and also it will be acting as a pick-up line during the time under consideration. The assistance of a computer will probably be required for a full analysis of cross-talk in the back-wiring panel for a system of any size. In practical cases where a computer (or the necessary programs) is not available, reasonable freedom from cross-talk can be ensured by running shorter tracks which will be unlikely to be switched simultaneously between the inevitable long tracks.

15.5.1 LOGICAL STATE OF BACK WIRING

One point which is worth noting is that less cross-talk will result if all 'dead' or 'waiting' tracks are held in the 'low' state.

When D.T.L. was the major logic family in use, it was customary to hold 'dead' tracks in the 'high' state, so that all control signals, data highways, etc. worked in the negative logic convention where a logical '1' is represented by a low voltage level. Negative logic convention was used for as much of the back wiring as possible so that the maximum possible use could be made of the distributed 'wired-OR' and to ensure a sharp leading edge to a control pulse. As explained in Section 14.2.4, a distributed 'wired-OR' may cause undesirable transients and its use on high speed systems is of questionable value. Most devices in the T.T.L. ranges have an active pull-up circuit which precludes the use of the 'wired-OR', and gives sharp rising edges, so there can be no valid reason for continuing to design systems with the back wiring in the negative logic convention.

In most digital systems, it should be possible to 'design-out' most of the worst effects of cross-talk by suitable grouping of the tracks on the back wiring. Signals on many of the lines will probably be strobed, and if cross-talk from the strobe signals presents a problem, these signal lines can usually be specially screened from the rest of the wiring.

15.6 Cross-talk in Discrete Wiring

When printed circuit back wiring first became a viable commercial proposition, some engineers objected to its use on the grounds that the necessary orderly arrangement of tracks would cause quite intolerable cross-talk.

Various efforts have been made to discover just how much cross-talk can occur in discrete wiring. In the most comprehensive series of tests known to the author, sockets with their pins on 0·150 in centres were interconnected by 7/0076 P.T.F.E. insulated wires, with pairs of wires run between pairs of pins on the connectors. The connectors were mounted on a copper-clad fibreglass panel which formed an earth plane for the system, and the wires were laid as close as possible to this panel but not secured to it. The main conclusion derived from this set of experiments was that cross-talk in discrete wiring is unpredictable and inconsistent unless all the wires are tightly bound into cableforms, in which case the cross-talk can be expected to be more consistent, almost unpredictable within economic design times, and probably intolerable on wires over 12 in long! Although the results obtained from the discrete wire system varied every time the system was touched, the lowest values measured were little better than on the printed boards with tracks on 0·050 in centres, and some values were up to 20 per cent higher. The indisputable fact that many successful systems have been built with their back wiring in discrete form is a convincing demonstration that in the great majority of digital systems it is possible to 'strobe out' cross-talk and other undesirable effects, and even in very large systems there may be only a dozen or so lines on which cross-talk can be said to be really critical.

15.7 Critical Length of Pick-up Lines

In Section 14.4.3.4 it was explained that ringing on the signal line will cause multiple cross-talk signals on the pick-up line, and to find the total cross-talk these multiple signals must be evaluated and added. Provided that the signal line is shorter than one and a half times the critical length, the tedious process of evaluating the cross-talk for all the reflections can be eliminated by regarding only the first step of a falling edge as significant, and only the first two steps of a rising edge as significant. (See Figs. 14.11–14.14.) Cross-talk from a falling edge is evaluated as described, but for the rising edge it can be assumed that the effect of the ring on the signal line is to slow down the rising edge such that for a signal line of critical length the edge speed of the driving gate is halved, and for a line one and a half times the critical length, the effective edge speed will be one third of the true speed of the gate. For evaluating the pick-up, this lower effective edge speed of the driving gate will increase the critical length of the pick-up line and the cross-talk is simply evaluated for the slower rising edge with all reflections ignored.

16

Printed Circuit Board Design for T.T.L.

16.1 Mounting of T.T.L. Devices

16.1.1 BOARD LAYOUT

When the decision to build a system from T.T.L. has been taken, the first question to be asked is how will the devices be mounted? Board size is partly a mechanical problem, and partly a system problem (see Section 18.2.2.2). Board layout also has mechanical and electrical implications.

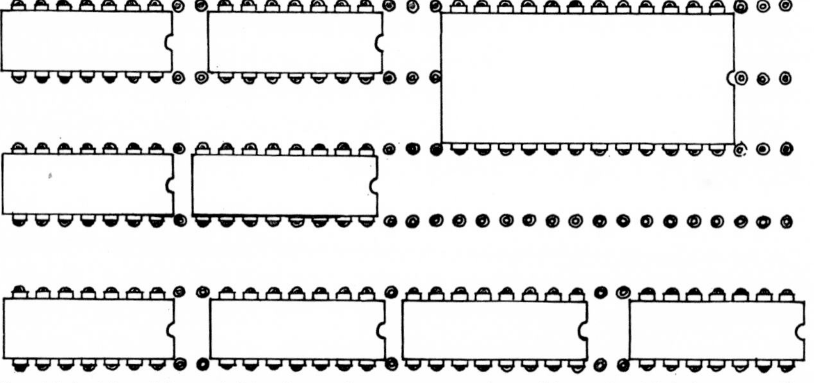

Fig. 16.1 14- , 16- and 24-pin packages on portion of board which has standard package-mounting holes in rows on 0·300 in centres.

The mechanical aspects are dealt with in *Printed Circuit Boards for Microelectronics*. From these, the form of board most likely to be adopted is the double-sided, X–Y co-ordinate layout, with the packages in uniform rows (see Fig. 15.1). If the larger M.S.I. Dual-in-line packages are to be used, board layout and standardization of artwork and drilling is simplified if the basic board design has rows of possible package mounting holes on 0·300 in centres, with each row comprising a continuous row of holes on 0·100 in centres. 14- or 16-pin packages are accommodated in adjacent rows of holes whereas 24-pin and larger packages can span any three rows (see Fig. 16.1). The end-to-end spacing of the devices can be adjusted to suit the particular circuit to be implemented on each board.

16.1.2 TRACK WIDTHS AND CHARACTERISTIC IMPEDANCE

The mechanical factors which govern the choice of track widths and pad diameters are discussed in *Printed Circuit Boards for Microelectronics*, and from these it can be seen that the track width chosen is almost certain to be between 0·005 and 0·020 in. From Fig. 14.1 it can be seen that the characteristic impedance will lie between 145 and 100 Ω on double-sided boards one sixteenth of an inch thick, or between 75 and 100 Ω on boards with internal earth planes. As can be seen from Figs. 14.11–14.16 this range of characteristic impedances is quite suitable for use with T.T.L. devices. If Series 2 devices are to be used, overshoot on rising edges will be minimized if the higher impedances are avoided (see Fig. 14.14). On the other hand, low impedance tracks can cause undesirably low steps at the driving end of a line.

One point which must be borne in mind when a track width is chosen is that a gate might have to drive two lines in parallel. If low impedance lines are used, gates which drive two or more lines in parallel will have several steps on their rising edges, and the dissipation of the gates will be increased. Line impedances of less than 100 Ω should therefore be avoided whenever possible.

The choice of 100 Ω for the line impedance has one practical benefit for the circuit designer—if the gate input and output characteristics are drawn at a scale of 1 V and 10 mA per unit, all construction lines for the determination of ring and the evaluation of cross-talk can be drawn in very quickly as they will all lie at 45 degrees to the axes of the drawing!

16.1.3 MAXIMUM TRACK LENGTH TO GUARANTEE FREEDOM FROM SPURIOUS SWITCHING CAUSED BY CROSS-TALK

The maximum distance for which tracks can be allowed to run parallel without a risk of spurious switching occurring because of cross-talk depends on the values of K_B and K_C, and on the voltage noise margins of the devices used. In cases where the cross-talk signals will be of brief duration (say less than 10 ns) the a.c. noise immunity of the gate may be considered (see Section 9.4.5).

16.1.3.1 *Noise margins*

As explained in Section 9.4.2, the worst-case voltage noise margin for T.T.L. is 0·35 V, whereas a practical value in a chain of gates could be 1·25 or 1·35 V. Worst-case design will result in a very short limit to the coupled length of tracks, but 'statistical' design can lead to occasional failures in test or commissioning. It must be the responsibility of the individual designer (or the leader of a design team) to decide whether to use what may be unduly restrictive worst-case calculations, or to decide on exactly what noise margin he will allow. The author's general practice is to allow for 1 V noise margins, which leaves a little in hand in most cases, and to ignore the a.c. noise immunity.

16.1.3.2 *Voltage swing*

As explained in Section 14.4.3.1, the worst pick-up voltage will be $2K_BE$, or if a multiple-line configuration is being considered, $2K_BK_CE$.

The value of E can be between a worst-case minimum of 2·0 V up to around or even over 4 V if overswing occurs on a Series 2 gate (see Section 14.3.2.1). Here again, it must be the responsibility of the design engineer to decide what value should be used; the author usually takes 3 V as a reasonable value.

16.1.3.3 *Values of* K_B

If we assume a 3 V swing and 1 V noise margin, we can say that $6K_BK_C$ must be less than 1. As explained in Chapter 15, K_C should not exceed 3, and with a little care in design it can be kept below 2. So for cases of a single pair of lines, the length can be unlimited if K_B is less than 0·16 or in multiple-line configurations if K_B is less than 0·05. From Fig. 14.18 it can be seen that with a spacing of 0·030 in between tracks on double-sided boards one sixteenth of an inch thick, K_B will be just under 0·15. A value of 0·05 will be possible only on very thin boards or on multi-layer boards.

16.1.3.4 *Cross-talk on double-sided boards*

If one sixteenth of an inch thick double-sided boards are used, cross-talk should cause no trouble if tracks 0·020 in wide, or less, on 0·050 in centres are used, and care is taken in the design to restrict K_C to less than 1. If K_C cannot be restricted then the track length will have to be restricted to such a fraction of the critical length as will reduce the cross-talk to a peak value of 1 V. For the track dimensions quoted above, an adequate approximation is to limit the coupled length of tracks to one third of the critical length when K_C is unlimited. In most practical cases, K_C is likely to be less than 2, so half the critical length can be allowed as a maximum coupled length.

16.1.3.5 *Edge speeds of T.T.L. and track length*

Edge speeds of T.T.L. devices can be taken as 8 ns for Series 1 devices and 2–3 ns for Series 2. For tracks 0·020 in wide on one sixteenth of an inch thick G10 board, these figures yield critical lengths of 27 in and 6·7–10 in. Thus if tracks are run on 0·050 in centres and K_C is unlimited, the coupled lengths of lines should not be allowed to exceed 9 in for Series 1 devices or 2·2–3·3 in for Series 2. If K_C can be limited to 2, these figures can be relaxed to 13·5 in or 3·3–5 in. (*Note:* two values have been quoted for Series 2 devices because the edge speeds of Series 9000 devices are generally not quite so fast as those of the other Series 2 families. Since the noise margin of the gates has been taken arbitrarily as 1 V, all these figures must be regarded as reasonably typical values and *not* worst-case limits—worst-case design would give impracticably short track lengths.)

16.1.3.6 *Effect of reducing clearance between tracks*

If the board layout is tightly packed, there may be a temptation to reduce the clearances between tracks to a minimum figure based on etching and soldering considerations. A possible value is 0·015 in, which will yield a K_B of 0·22. When $K_C = 1$, the peak value of cross-talk will be 1·32 V, so track coupled lengths would have to be reduced to 20·5 in for Series 1 devices and 5·1 in for Series 2. For any possible track configuration ($K_C = 3$), the lengths would be 6·8 in and 1·7 in. Such restrictions on track length represent quite a severe penalty to have to pay for squeezing in a few extra tracks, and it is well worth trying to restrict board layout to running all tracks on the lines of a 0·050 in grid. (This simplifies the layout as well as easing the cross-talk restrictions!)

It may be noted that the simplification mentioned in Section 15.7 of allowing twice the critical length for the pick-up line has been ignored. This is because the simplification applies only to rising-edge cross-talk.

The figures quoted above can be used for rapid assessment of a possible design, but designers are urged to calculate their own figures from values of edge speeds applicable to the devices they intend to use, and based on voltage noise margins selected with due regard to the environment in which the equipment will have to operate—there are sources of noise other than cross-talk!

16.1.4 OTHER LIMITS ON TRACK LENGTH

If the length of tracks is unlimited, care must be taken to calculate the timings of the system with due allowances for line delays. As explained in Section 14.3.2.1, a gate at the 'driving' end of a long line may not switch until after twice the line delay. If such calculations are to be avoided, the line lengths should be restricted. For tracks on double-sided boards, 12 in of track will load the gate output with about 18 pF, which is only slightly more than the figure at which gate delays and edge speeds are normally specified and tested. For Series 1 devices 12 in is just under half the critical length, which means that ringing is unlikely to cause any problems.

16.1.5 APPLICATION OF LIMITS TO TRACK LENGTH

Any such limits applied to track lengths should not be regarded as 'absolute'. Board designers should be asked to work within the limit whenever possible, and to report all cases in which it seems likely that the limit will be exceeded. These cases can then be examined to see whether the excess length is likely to have any adverse effect on the working of the system. The track concerned may not be in a critically timed portion of the circuit, or in the case of cross-talk, any spurious signals might occur only at times when they will be 'blocked' by an inhibited input to a register, etc. In the author's experience limits such as those quoted here (for Series 1 logic) are not exceeded very often, and when the limits are exceeded, it is very rare to find that a track is 'critical'.

A limit on track length will mean that there will also be a limit to the size of system which can be built without resorting to the use of line drivers, screened lines, etc., but if 12 in is taken as the limit on track length, quite large systems can be built. With a little care in the system design, systems with several thousand packages mounted on double-sided boards can be implemented without too many infringements of the 12 in limit.

When the board size is chosen, the maximum allowable track length can be borne in mind. If the length plus the width of the board is about an inch more than the maximum allowable track length, it is unlikely that any tracks within the boards will exceed the limit length if an X–Y co-ordinate layout with random via holes is used (but see Section 18.2.2.1). Such a limit to board size will not remove the need to check track lengths entirely—the lengths of track which run via the back wiring to other boards in the system will still have to be checked. In large systems these tracks are more likely to exceed the limit lengths than those tracks which are confined to one board, and it may well be better to use large boards which will minimize the length of the back wiring than to attempt to apply any 'automatic' limit to track lengths within the boards.

16.2 Power Distribution and Voltage Spike

16.2.1 SUPPLY RAIL IMPEDANCE

The lowest possible supply rail impedance will be achieved when the packages are mounted on multi-layer printed circuit boards which have complete power and earth planes with the minimum thickness practicable of dielectric between them. Such boards would probably be made up from two double-sided boards, each 0·025 in thick, with one signal layer and one plane each. The planes would be separated by about 0·006 in of 'B-stage' when the two halves of the board are bonded together. Such multi-layer boards are considerably dearer than double-sided boards, and since all signal tracks are on the board surfaces, the multi-layer boards will be at the best only slightly smaller than would double-sided boards for the same circuits. Thus it is unlikely that internal-plane boards will be used if double-sided boards can be used without risk of spurious switching caused by the rail spike.

16.2.1.1 *Power supply on double-sided boards*

On double-sided boards the distribution of power and earth to all packages can be arranged in a variety of ways (see *Printed Circuit Boards for Microelectronics*). Whatever system is used, the best results will be achieved if both power and earth rails form continuous closed 'grids', and the author has adopted the practice of allocating the end pins of the board connector for earth, and the next pins in for the power rail. On each board earth and power 'bus tracks' are run along the edges of the board, and pairs of tracks which run across the major axes of the packages distribute power and earth from these 'bus tracks' to each device. In the middle of each board, all power and earth tracks are connected by links or straps as shown in Fig. 16.2. Thus there is always a multiplicity of power and earth tracks between any two packages on the board.

16.2.1.2 *Calculated supply rail impedance*

On boards laid out with tracks 0·020 in wide on 0·050 in centres and where the packages are placed on 0·6 in centres side-to-side as suggested in Section 16.1, the self-inductance of the length of track between two packages will be about 15 nanohenries (nH). Adding allowances for package pins, lead frame, and bonding wires, the chip-to-chip inductance will be about 20 nH. Gate chips have a supply rail to earth capacitance of around 50 pF, and flip-flops and M.S.I. devices have larger capacitances. For the purposes of supply rail calculations, the 'worse than worst-case' value of 30 pF has been used.

Thus each section of the supply rails across the board will have inductances and capacitances as shown in Fig. 16.3 provided that all package positions on the board are filled. The calculated characteristic impedance of such a line is 36 Ω.

Devices a quarter of the way in from the edge of the board are therefore connected to two 36 Ω lines in parallel (Fig. 16.4(a)) whereas devices at the centre or edges of the board are connected to at least one 36 Ω line and two further lines of slightly higher impedance (Fig. 16.4 (b) and (c)). Thus the highest possible source impedance for the power supply to any device is 18 Ω.

Calculations based on a value of 18 Ω are pessimistic because the effects of resistance have been ignored. Resistances in the chips and tracks lower

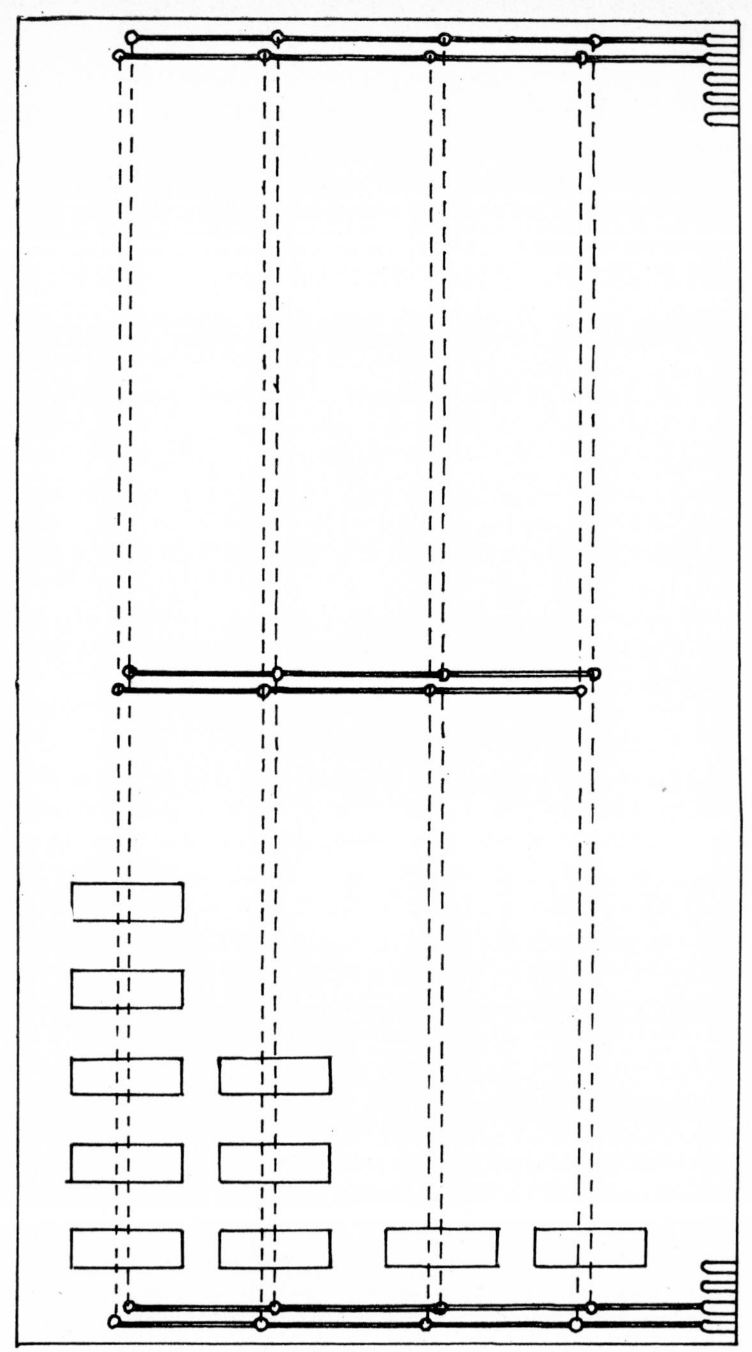

Fig. 16.2 Typical power-distribution grid. (The approximate route of tracks on the back of the board is shown by a dotted line. In practice they would bend to contact or pass between package pins. For devices with power and earth on pins 14 and 7 the tracks would run past the ends of the packages.)

the line impedances and also 'damp' the lines. The capacitance of the tracks has been ignored, as has the fact that on some T.T.L. devices the power and earth tracks can be run closely adjacent over the board.

Calculations on a line with 40 nH inductance and 30 pF capacitance yield a stage delay of 1·1 ns, and a resonant frequency of about 150 MHz. Thus voltage spikes on the line, caused by the T.T.L. current spikes, can be expected to be about 7 ns wide.

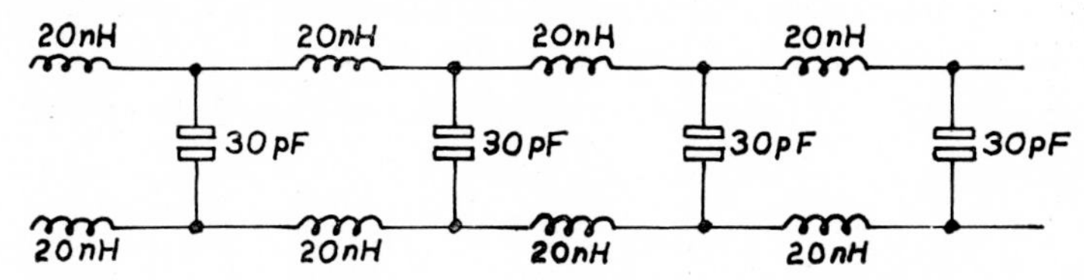

Fig. 16.3 Equivalent inductance and capacitance of power distribution line.

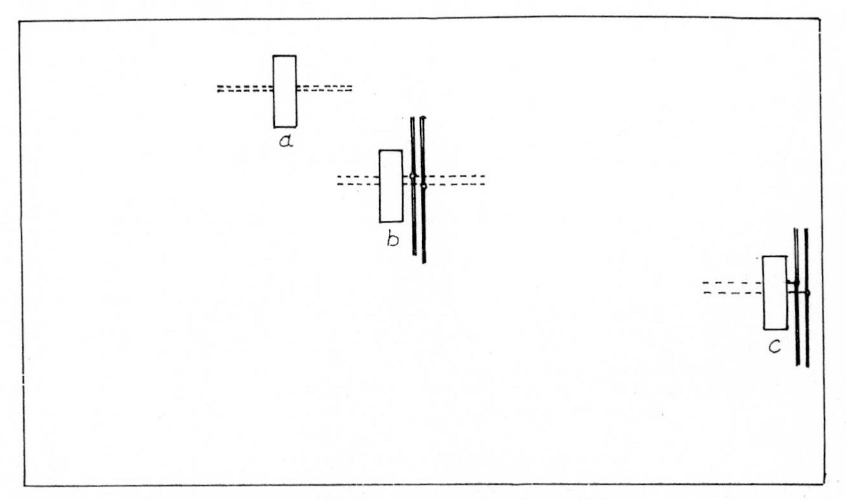

Fig. 16.4 Parallel power-supply lines to devices.

16.2.2 VOLTAGE SPIKES ON TYPICAL BOARDS

16.2.2.1 *Maximum amplitude of current spike*

The T.T.L. device specifications place no limit on the amplitude of the current spike, so no true 'worst-case' calculation can be done. Measurements have shown that typically the amplitude of the current spike is about one third of the short-circuit output current of the device, provided that the output node is not heavily capacitively loaded (see Fig. 10.3). Since the output short-circuit current I_{OS} is specified as 55 mA for Series 1 devices (57 mA for some M.S.I. devices and flip-flops) and 100 mA maximum for Series 2 devices, it can be assumed safely that the current spikes will not exceed 18 mA and 33 mA. (Measurements of current spikes have indicated that these limits are, if anything, pessimistic.)

16.2.2.2 *Worst-case voltage spike*

A current spike of 18 mA with a source impedance of 18 Ω will result in a worst-case voltage spike of 324 mV for Series 1 devices, and the 33 mA current spike of a Series 2 device produces a voltage spike of 595 mV. Measurements taken on a 13 board logic system, which had an average of about 44 packages per board and a power distribution system as recommended, showed that these figures can indeed be regarded as 'worse than

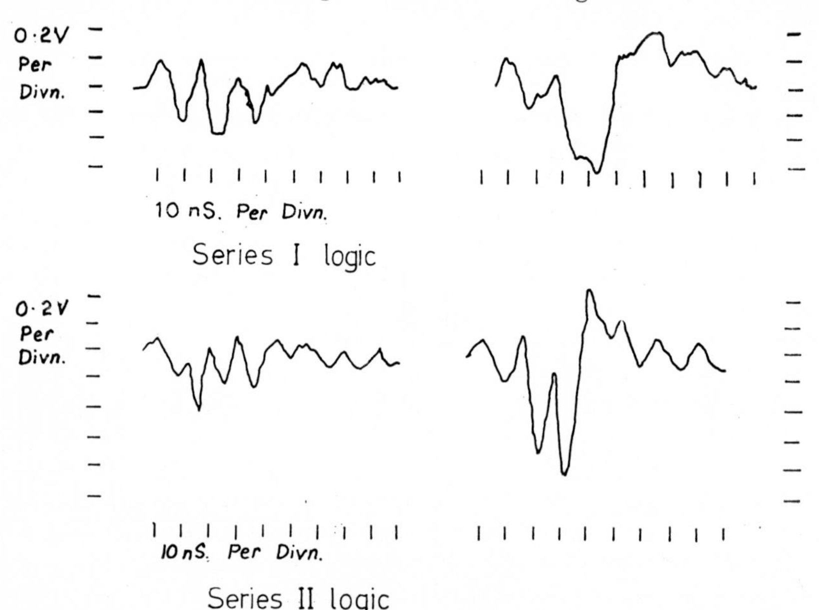

Fig. 16.5 Worst-power rail voltage spikes measured on undecoupled double-sided boards (49 packages).

worst case'. The whole system was completely undecoupled, and contained a mixture of Series 1 and Series 2 devices from four different manufacturers. During the tests done on this system, up to 48 gates were switched simultaneously, of which 12 were on the same board, and 11 were in the same row on the board. Generally the rail spikes measured were less than 200 mV; about 90 per cent were always less than 300 mV, and a few spikes were found around 1 V amplitude. (It must be borne in mind that this 1 V spike was found on an undecoupled system built specially to evaluate such items as rail spikes, and was the result of a number of Series 2 devices switching simultaneously. For figures on practical production boards, see Section 16.3.3.) Fig. 16.5 shows the worst rail spikes found in the system, two from a board of Series 1 logic, and two from a board of Series 2. Tables 16.1 and 16.2 give plots of the voltage spikes measured on each package on two boards, measured with the boards taken out of the system and fed with the input pattern which has been found to generate the worst voltage spikes on the supply rail.

16.2.3 EFFECT OF THE VOLTAGE SPIKE ON THE OUTPUT SIGNAL

If the boards are designed to withstand cross-talk caused by the full logic swing of the gate, track to track coupling of the supply rail voltage spike onto a logic signal track can be ignored. As well as the coupling between adjacent tracks, coupling within the device which generates the rail spike must also be considered. Such coupling can exist between the supply rail and the output of the gate which is generating the spike, and it can also exist between the supply rail and another, unswitched, output on the same chip as the output which is causing the spike.

In the case of coupling to an unswitched output, this output can be ON or OFF. If it is ON, the output will be clamped to earth by the main output transistor VT5, whereas, since VT4 is OFF, the coupling from the rail will be through the stray capacitance associated with VT4, which, as explained in Section 3.1.4, appears to be about 4 pF. At the operating frequencies of the gate this capacitance represents over 100 Ω impedance, and so any disturbance on the output will be less than 10 per cent of the voltage spike on the rail, or less than 33 mV for Series 1 devices and less than 60 mV for Series 2.

When the gate is OFF, VT4 is conducting, acting as an unsaturated emitter follower. This will not allow any appreciable direct coupling of the rail spike through VT4, and, as explained in Section 9.4.3, the possible path through R2 and D3 (or VT3) can be ignored. This leaves only direct coupling through the stray or parasitic capacitances. As explained in Section 3.1.4, these capacitances cannot be measured easily, but direct measurements of the coupling of specially generated pulses showed that the worst coupling that can be expected is 50 per cent; i.e., 163 mV on Series 1 devices, and 300 mV on Series 2 (if the voltage spikes were at the 'worse than worst-case' limits calculated).

During a switching transition (turn-off) VT4 may be saturated, with its current controlled by R4. On Series 1 devices R4 usually has a value of 150 Ω, and on Series 2 devices it is usually 80 Ω. During the turn-off transition, VT3 has just desaturated, and its equivalent resistance is indeterminate. Initially the equivalent resistance between the output terminal and earth will be less than 80 Ω, so the breakthrough of the voltage spike to the output will be less than 50 per cent. As the output potential rises, the breakthrough will increase until VT4 desaturates. Such coupling of the

TABLE 16.1 Turn OFF Voltage Spikes (in millivolts) on undecoupled board of Series 1 Logic, measured against Earth Level at edge-connector.

	EDGE-CONNECTOR										
1. Peak +ve	210	190	200	200	210	200	200	180	160	150	130
Peak −ve	220	230	230	230	210	200	180	200	240	270	210
2. Peak +ve	200	190	170	170	190	200	200	210	210	210	180
Peak −ve	210	210	230	240	240	270	270	270	270	250	260
3. Peak +ve	240	250	240	260	250	250	240	250	230	220	190
Peak −ve	340	350	340	350	360	350	360	350	340	310	300
4. Peak +ve	260	280	300	320		340	360	360	370	380	380
Peak −ve	430	440	430	440		550	600	660	700	730	740
5. Peak +ve	240	210	210	200			230			220	
Peak −ve	210	440	430	400			360			340	

TABLE 16.2 Turn OFF Voltage Spikes (in millivolts) on undecoupled board of Series 2 Logic, measured against Earth Level at edge-connector

	EDGE-CONNECTOR										
1. Peak +ve	230	240	240	310	330	340	340	350	330	320	280
Peak −ve	300	310	290	350	330	330	320	330	260	260	330
2. Peak +ve	250	200	280	300	300	300	290	310	290	290	290
Peak −ve	300	370	420	460	500	530	530	460	410	350	290
3. Peak +ve	290	320	340	340	320	310	310	310	270	270	270
Peak −ve	470	480	450	480	480	460	420	380	320	300	240
4. Peak +ve	310	360	390	430		510	530	580	600	640	660
Peak −ve	650	690	740	770		780	820	810	840	850	860
5. Peak +ve	410	370	360	360			290			290	
Peak −ve	450	430	390	380			310			290	

voltage spike to the output of a gate which is turning off can cause discontinuities in the rising edge of the output pulse, but although such discontinuities have been observed at or near the input threshold level of a following gate, the discontinuities have always been too small and too fast to cause any spurious switching of the following gate.

Coupling of the small 'turn-on spike' to the output of the device which is turning on can be ignored.

Summarizing, the worst-case disturbances which can be expected from a single rail spike are 33 mV on a '0' level output and 162 mV on the '1' level for Series 1 devices, and 60 mV and 300 mV for Series 2.

16.2.4 ADDITION OF VOLTAGE SPIKES CAUSED BY GATES SWITCHING SIMULTANEOUSLY

When two gates on the same chip are switched simultaneously the voltage spikes will generally be added to give approximately double the figures quoted above, because by virtue of their simultaneous diffusion the two gates can generally be expected to have very similar properties and with simultaneous inputs the current spikes will occur within a nanosecond or two of one another. However, when gates in different packages on the same power distribution line are fed with simultaneous signals, it will generally be found that the resulting voltage spike on the rail will be less than the sum of the individual spikes because the delay along the power distribution lines will add to the separation in time of the spikes likely to be present due to differences in the delays of the gates.

16.2.5 EFFECT OF EARTH RAIL IMPEDANCE AND SPIKES ON THE BOARD EARTH

The current spike has to flow through the earth supply rail as well as the power supply rail, and since on boards laid out as described in Section 16.2.1.1, the earth rail is similar to the power supply rail, it might be expected that on such boards a 300 mV voltage spike would appear as a 150 mV dip of the power rail and a 150 mV rise on the earth rail. Such an effect would be equivalent to a 150 mV disturbance on a signal line, since the switching threshold is determined with reference to the earth rail level. However, measurements on working boards have shown that this does not occur. If measurements are taken with respect to earth at the corner of a board, gradients and spikes on the earth lines at other points over the board can be seen and measured. (On one board, up to 300 mV spikes were measured.) Measurements from package earth pins to input pins on the same packages have demonstrated that these potential gradients are general over the whole board, and the inputs to gates do not 'see' the spikes suggested above.

Accurate measurements of small spikes on earth rails are extremely difficult to take—it is virtually impossible to find a 'solid' reference point, and to eliminate all stray pick-up from the oscilloscope probes, but there can be no doubt that the power and earth distribution described above will ensure trouble-free working of T.T.L. devices in all normal logic applications. (A whole board of line drivers all switched simultaneously could be expected to cause trouble with almost any power distribution system!)

16.3 Decoupling

16.3.1 INTRODUCTION

From the foregoing it should be clear that the rail spike is unlikely to cause malfunctioning in reasonably well laid out T.T.L. systems. In fact it is quite possible to run systems built from Series 2 devices without any decoupling other than that given by the presence of other devices on the boards.

If analogue devices are to be driven from the same supply rails as the logic devices, then special arrangements for decoupling will have to be made, but on purely logic boards there is no merit in spending money on ensuring a 'clean' supply rail. As has been shown, a spike on the rail will not couple fully onto an output signal, and since the first positive going noise signal will drive a '1' level signal to a higher than normal voltage level, much of the coupling of rail spikes onto signal lines can be ignored, because although the line will appear 'noisy', it is probable that the lowest voltage reached by the line will be higher than the specified minimum '1' level. Spikes of up to 500 mV on a supply rail can be tolerated, and even larger spikes will usually prove to be acceptable even under worst-case environmental conditions of temperature, etc., unless there are external sources of noise which could affect the logic unit under consideration.

16.3.2 THE USE OF INTERNAL POWER AND EARTH PLANES

Boards for mounting T.T.L. devices can be made as four-layer boards with signal tracks on the surfaces and internal power and earth planes as described in Section 16.2.1. The inter-plane capacitance of such boards may be insufficient to act as a 'tank' capacitor, and it may be advisable to mount capacitors (such as a pair of 2·2 microfarad (μF) tantalum capacitors) in order to keep a fairly smooth supply rail throughout the system.

In an attempt to study the effects of internal planes on the working of Series 2 T.T.L. without any risk of confusion caused by differences in system design, track widths, etc., the artwork for an existing 49-package D.I.P. board (double-sided, X–Y co-ordinate layout with 0·020 in wide tracks, on one sixteenth of an inch G10 material) was taken and all power and earth tracks were stripped off. Internal power and earth planes were

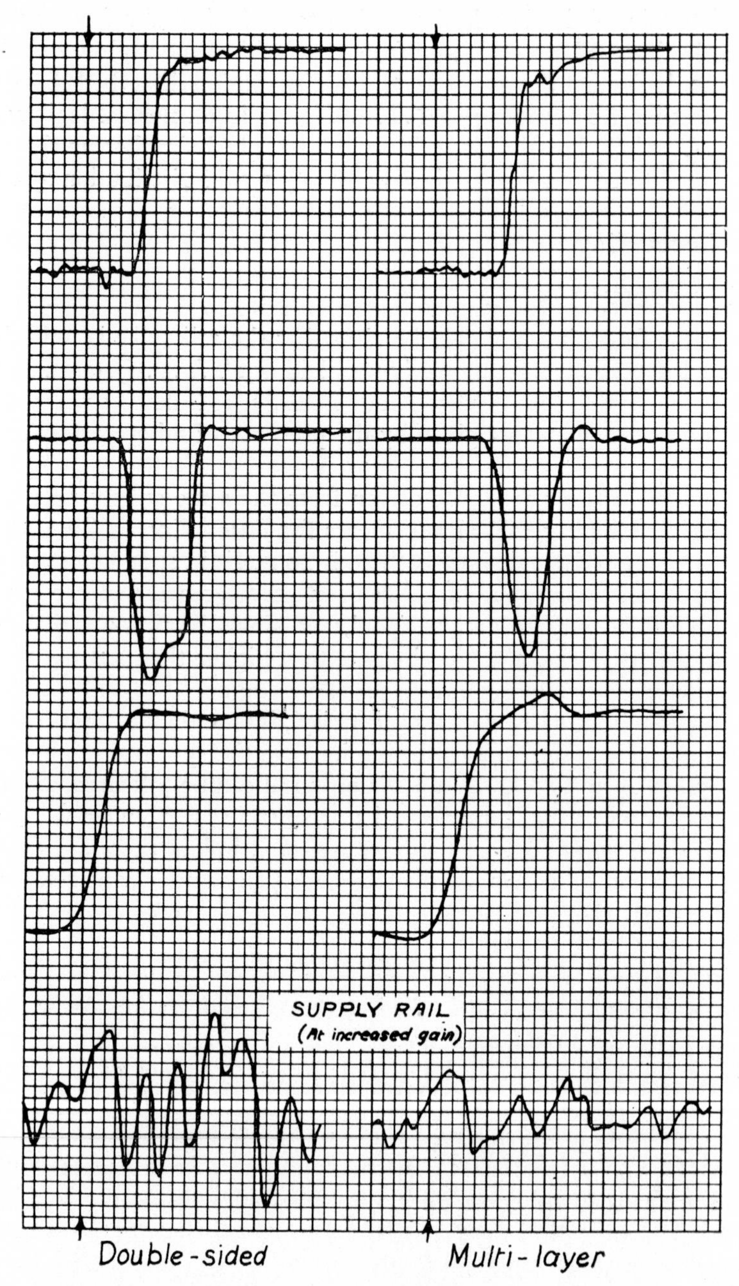

Timing references arrowed

Fig. 16.6 Waveforms measured at corresponding points on double-sided and multi-layer boards.

laid-up, and a four-layer board was made, in which the internal planes were separated by 0·008 in. A further experiment was also conducted, in which the power and earth wiring was replaced on the original artwork and all areas of the board surfaces not occupied by tracks were covered with partial earth planes.

An original double-sided board, the double-sided board with extra earth area added, and the four-layer board were packaged with Series 2 devices which had been very carefully selected for the best possible match on all parameters. The boards were then run in turn in an experimental 13 board system, with an independent d.c. power supply to the board under test, and rail spikes, ring, and cross-talk within the board circuit were measured and recorded. No difference at all could be detected in the performance of the two double-sided boards, so the board with the partial earth planes on its surfaces was discarded.

The multi-layer board showed no measurable improvement in cross-talk (of which there was little on either board), but the effects of line ringing were more marked on the multi-layer board than on the double-sided. Fig. 16.6 shows some of the waveforms measured on the boards. No scales are quoted on the drawings because the amplifiers (X and Y) on the sampling oscilloscope used were run uncalibrated so as to give the largest picture possible. Pairs of waveforms were, of course, measured on the same settings.

As expected, differential mode measurements between power and earth pins on the packages showed that there was less noise on the multi-layer board than on the double-sided. On the multi-layer board the peak-to-peak swing on the supply rail was just over half that measured on the double-sided board.

However, the tests revealed a very significant difference in performance between the boards. The signal delays on the multi-layer board were all around 10 per cent greater than those on the double-sided board. Since the only reason for adopting Series 2 T.T.L. instead of Series 1 was to achieve substantially faster working, it was considered that the use of multi-layer boards could not be justified as an economic proposition. Provided that, when parts of the logic system are allocated to boards and the boards are laid out, reasonable care is taken to separate any large groups of devices which will switch simultaneously there should be no difficulty in providing an adequate power supply on double-sided boards, and money spent on making the more expensive multi-layer boards will be money wasted.

16.3.3 PRACTICAL DECOUPLING ON DOUBLE-SIDED BOARDS

Although double-sided boards can be run without any decoupling other than that offered by the devices themselves, it is advisable to add some decoupling capacitors as a safety margin against noise generated externally to the board. On boards laid out as described in Section 16.2.1.1, the obvious places to mount capacitors are the four corners of the power and earth grid and the ends of the centre pair of 'straps' across the grid. Experiments have shown that on a 44-package board, no advantages are gained by using capacitors larger than 0·01 μF, and the Tropyfol F type gave the best results of the types tried (as well as being one of the cheapest).

If it is essential to lay out a board with a row of packages which will all be switched simultaneously, it might be necessary to add extra decoupling capacitors along this row. Generally, better results will be obtained by arranging the layout such that other devices are interspersed along the row of simultaneously switching packages.

When all the signal tracks have been laid down on a board, it is excellent practice to add extra pairs of 'straps' across the power and earth tracks midway between the ends of the boards and the recommended pair of 'straps' in the middle of the board, if there is sufficient space in the track layout.

On boards laid out in this way, carrying up to 48 Series 1 packages and 6 Tropyfol F capacitors, no rail spikes in excess of 300 mV peak to peak have been found, even when several gates in a row are switching together.

17

Practical Considerations in the use of T.T.L.

17.1 Power Supply

17.1.1 NORMAL LIMITS

The normal rail potential for T.T.L. is +5 V. Generally, military devices are specified as working reliably between 4·5 and 5·5 V and commercial devices between 4·75 and 5·25 V. In fact all devices the author has used worked reliably down to considerably lower voltages. The level of the input threshold is not affected by variations in rail voltage, but the '1' level output voltage varies directly with the rail voltage.

Some T.T.L. devices are specified as being able to withstand a surge of 12 V for 1 s on the supply terminal, but this can be applied only if there is no other device connected to the output of the device to which the surge is applied. If two or more gates are connected in cascade the surge should be limited to less than 7 V, as otherwise a 'high' output (approximately 1·5 V below the supply rail) can apply more than 5·5 V to the input of the next gate in the chain. An experimental system built by the author has twice withstood power supply regulation failures which left the supply at about $7\frac{1}{2}$ V for several hours, but this treatment cannot be recommended!

If the normal power supply connection is taken negative with respect to earth, excessive currents can flow at potentials beyond about −3 V. Packages may be damaged, if not destroyed, if the supply and earth connections to a board are inadvertantly inverted.

17.1.2 PERFORMANCE WHILE SUPPLY RAIL IS RISING OR FALLING

A gate with no power supply connected will pass only very small currents (less than 50 μA) when up to 5 V is applied to an input or an output terminal. Increases in applied voltages up to about 5·7 V may cause breakdown in the output stage or the input. (Note that positive voltages should never be applied to the output of a gate which may be

ON, other than through a series resistor of sufficiently high value to limit the current into the ON output to less than about 20 mA.)

As the power supply voltage is increased from 0 V to 5 V, the output of a gate with its inputs open circuited and its output connected to the input of a 'live' gate falls to the normal 'low' level as the supply potential passes through about 2·5 V. The output of a gate with its inputs earthed, which feeds into a 'live' gate, sits at the 'uncommitted input' level of the driven ('live') gate until the supply voltage reaches about 3 V, then the output rises, following about 1·5 V below the rising supply voltage, up to the normal '1' level. At no time during the turning-on or turning-off of the supply does the output of a gate with its inputs held low sink any current. However, if the rise in voltage of the power supply is irregular, a gate with its inputs held 'high' or open circuit could sink current intermittently in its output while the supply rail is at around 2·5 V (see Section 9.5.3).

If any portion of an equipment is intended to have its power switched off while the remainder of the equipment is working, all logic outputs from this portion of the equipment should be designed to be at a 'high' level before the power supply is removed. There will then be no disturbance to the working of the rest of the equipment.

17.2 Connections to the Outputs of T.T.L. Gates

17.2.1 INTERCONNECTION OF GATE OUTPUTS

Normal T.T.L. gate outputs cannot be interconnected to form a 'wired-OR' function because if one gate is ON while the other is OFF the OFF gate output will source almost all its short-circuit output current, and the ON gate will have its output voltage raised through sinking the excessive current, which may pull it up out of saturation. Thus the dissipation in the two gates will be increased and low level noise immunity will be lost. 'Open collector' T.T.L. devices are available, as are 'tri-state output' devices, and these can have their outputs interconnected to form a wired-OR when necessary (but see Section 14.2.4). The use of open-collector devices and the wired-OR may sometimes be considered, but it should be borne in mind that fault diagnosis to a gate is almost impossible in wired-OR configurations since a short to earth or a faulty 'low' level output in any of the gates which drive the node will hold all the other outputs 'low'.

17.2.2 CONNECTION OF GATES IN PARALLEL TO DRIVE HIGH FAN-OUT NODES

When it is necessary to drive a fan-out higher than that quoted for a single gate, the driving gate will have to be duplicated or a line driver or buffer must be used. Line drivers are available only in the form of NAND Gates, so for any other logical configuration the driving gate will have to be duplicated. It is recommended that when this is done the total fan-out should be divided as evenly as possible between the two driving gates and each gate should feed its loads independently. This can result in timing differences between what should be simultaneous signals, and it can also result in unacceptable duplication of board connector pins. If either of these disadvantages is considered to be serious the outputs of the driving gates may be connected together to form a single node provided that their inputs are also connected in parallel. Such parallel connection of the two driving devices will ensure uniformity of timing, but it will often result in an increase in the duration of the switching spike in the supply current to the two gates, caused by differences in the delays in the gates. The user must decide whether the larger current spike, which will probably be larger both in amplitude and duration, is preferable to the difference in timing between the two outputs, which can usually be expected to be less than 4 or 5 ns, even if no precautions, such as selecting matched devices, are taken.

Such splitting down of large fan-outs to two independent nodes reduces the line currents, and it also simplifies fault diagnosis. Board layout is simplified if the determination of which devices are to be fed by each of the driving gates is made as the board is laid out. The person responsible for the board layout can be given a list of all the devices to be driven by the duplicate nodes, with an instruction to arrange the split so as to minimize the track length and equalize the fan-out on each node.

17.2.3 SHORTING OF INPUTS AND OUTPUTS DURING TESTING

During commissioning or testing of a board it may be necessary to earth the inputs to inverting gates to set their outputs 'high' in order to check the working of part of the circuit. If non-inverting gates are used, an input would be earthed to set the output 'low'. Any point in the middle of a circuit can be earthed to 'set' the outputs of all gates driven by the node. As described in Section 4.2.2, if the gate driving the node is in the 'high' state, it will source its full output short-circuit current, which will increase its dissipation by about 100 to 400 mW. If a multiplicity of such earthed nodes is used during testing, efforts should be made to ensure that no package on the board has more than one 'high' level output shorted to earth at any time.

It is not necessary to connect an open-circuit input to an artificial 'high' level source to set the output of an inverting gate 'low'—no current can flow out of the input terminal if it is left open circuit. If it is desired to 'set' a node in the middle of the circuit to a 'low' level, this should be done by earthing inputs to all the gates which feed into the gate whose output is to be set 'low'.

When a board is being tested, it might happen that there is a long track from the board connector to a gate input and for correct working of the board this track should be set at a 'high' level. If the track is left open circuit at the connector, it might pick up sufficient noise to disturb the correct working of the board, and so it is considered desirable to 'set' the track from a low impedance d.c. source. If this is done, it is better to use an independent power supply unit set to about 4·5 V than to use the 5 V supply rail to the board, because the supply rail may have surges or spikes which could damage the input stage of the gate.

17.2.4 CHECKING AND TESTING T.T.L. DEVICES ON BOARDS

When a fault is found on a printed circuit board which has been in service, it may be assumed to have failed because an integrated circuit has 'died'. However, before a repair is attempted it is advisable to check that the fault *is* in a device, and has not been caused by a badly soldered joint or a through-plated hole which has failed under environmental stress. Another possible failure mechanism is a short circuit caused by a 'whisker' of gold or tin–lead plate which has become detached from the edge of a track on a through-hole-plated board. These 'whiskers' should not occur on well-made boards—the use of tin–lead finish with a hot-oil dip to reflow the surface will eliminate any risk—but if they do occur, the faults they cause can be extremely difficult to diagnose since the short circuit may not be between adjacent tracks, and ordinary logical fault diagnosis procedures may not locate the cause of the trouble.

17.2.4.1 *Open circuit faults*

Open circuit faults on the inputs of a device are easy to locate—the output of the gate will fail to follow the input signals, or in the case of a flip-flop or M.S.I. element, the truth table will not be followed correctly. However, an open-circuit output will be held at either a 'low' level or at between 1 and 2 V by the input stages of the gates which are driven by the node, and the output may even appear to switch at about the right time if other inputs to the gates fed by the node are switched. The easiest way of resolving any doubts in such cases is to see the effect of connecting the node to earth and then to the 5 V rail via a resistor of about 220 Ω. The voltage on the node will fail to follow the correct output characteristic if there is a fault in the output stage.

17.2.4.2 *Short-circuit faults*

Fault conditions can arise in which a node will remain permanently at earth or at the supply rail level. When this occurs it is not possible to determine whether the fault is in the device which drives the node, or in one of the driven devices. The possible causes of such faults are a broken bond inside a package which allows the bond wire to touch the earth or power wire near the chip, or catastrophic breakdown of one of the semiconductor devices on the chip. When such a fault is found, and it is certain that the fault is in a device and not in the interconnection on the board, the node should be shorted to earth if it is sitting at a 'high' level, or if it is sitting at a 'low' level it should be shorted to 4·5 V after earthing one of the inputs to the gate which drives the node, or in the case of an M.S.I. device, after ensuring that its output should be at a 'high' level. This shorting will put 4·5 or 5 V across the 0·0013 in diameter aluminium or gold wire inside the faulty package, and the wire will promptly melt, which will convert the original short-circuit fault into an open-circuit fault, which is easily found. If the original short-circuit fault was a connection to the 5 volt rail in a 'driven' device, it is possible that the output stage which drives the node will have been damaged. This can be checked by replacing the now open-circuit faulty driven device; checking that all driven devices respond correctly when the node is shorted to earth, then checking the operation of the device which drives the node.

17.2.4.3 *Parametric degradations*

Parametric degradations in integrated circuits can be detected by 'pulling' the output up or down with two different value resistors to earth or to the supply rail. A 220 Ω and a 100 Ω resistor should serve to show whether the high and low level output characteristics of a device are normal. Various 'clip-on' testers which can be slipped over the body of a Dual-in-line package are available. These testers make electrical contact to the 'shoulders' of the D.I.P. leads and either indicate the logic levels of the pins, or, on some of the more sophisticated programmable testers, give a 'GO—NO-GO' indication. With the aid of one of these testers, or with a few resistors and shorting links (plus a meter, oscilloscope, or other normal test equipment) it is possible to locate faulty T.T.L. devices on a board without having to 'open' any of the connections, unless open-collector devices have been connected in a 'wired-OR' configuration.

One great advantage of such 'on the board' testing is that it eliminates all doubts as to whether a fault found in a device was the original cause of the board failing to work, or whether the fault was caused during the removal of the suspected package from the board. It has been the author's experience that 'on the board' testing can not only locate a faulty device but can also provide many clues as to why the fault occurred.

T.T.L. devices have not been in use long enough for there to be any significant 'end-of-life' information. Most faults experienced are early life failures, which could be virtually eliminated by burn-in, or are caused by 'sillies' such as excessive surges on power rails or the shorting of 'low' level outputs to high voltages.

17.3 Line Driving and Terminations

In some cases it may be necessary to drive electrically long lines, and because of the presence of externally generated noise, the line reflections may be considered an embarrassment. As explained in Chapter 14, the ringing can be eliminated by terminating the line in its characteristic impedance. If a series matching resistor is used at the ending end of the line, voltage noise margins will be reduced by the drop across the resistor caused by the load current, and the end result may be less overall noise immunity than if the line were left unterminated. A shunt termination at the receiving end of the line will eliminate the really dangerous reflection at the receiving end during turn-on, but it will not remove the uncertainty in the time of switching of a gate fed from the sending end of the line. (See Section 14.3.2.)

17.3.1 SHUNT TERMINATING NETWORKS

The only problem with a shunt terminating network is that the device which drives the line must be capable of handling the steady-state current in the termination as well as the normal load current of the devices which are driven from the line. No T.T.L. device can hold a 'high' level output into a load of around 100 Ω to earth, nor can any standard gate or flip-flop sink the 50 mA which would be necessary in the 'low' state if a single termination is taken to the 5 V rail.

Therefore a terminating network must be used. The requirements are that it must not tend to hold the line at any voltage between the normal 'high' and 'low' levels; that when it is pulled 'low' the driving gate must not be called onto sink more than about 40 mA (assuming that a line driver or buffer will be used to drive the line—for a normal gate the limit would have to be about 13 mA), and when the line is in the 'high' state the driving device must not be called on to source more than about 1 mA. Obviously for the third requirement to be met the 'rest' state of the terminating network must be above the normal T.T.L. 'high' level—i.e. more than about 3·5 V.

These requirements are met if the terminating network comprises a 470 Ω resistor to earth and a 120 Ω or 150 Ω resistor to the 5 V rail. The 120 Ω resistor gives a mean impedance of just over 95 Ω, whereas the 150 Ω resistor gives a mean of 112 Ω. Neither of these will pefectly terminate a 100 Ω line, but it must be borne in mind that the normal load on the line will appear in parallel with the terminating network. Also, the use of the 150 Ω resistor for the upper arm will result in less current for the driving element to sink in the 'low' state, and hence a higher fan-out from the line.

At 0·4 V on the line, with a high (5·5 V) supply rail, a worst-case arrangement of 5 per cent end-of-life resistors will source a current of just under 35 mA. This allows a fan-out of 8 Series 1 gates to be driven by a Series 74 buffer.

17.3.2 PARTIAL TERMINATION OF LINES DRIVEN BY STANDARD T.T.L. GATES

As a general rule, partial termination of a line driven by an ordinary T.T.L. gate is unlikely to have any significant effect on ringing. The overall effect will be similar to that of loading the line with the full specified fan-out of the driving gate, and, as can be seen from Fig. 14.15, the positive ring at the receiving end can be substantially the same for low and high fan-outs. Thus it is probable that the only practical difference a partial termination will make is to increase the dissipation of the unit—which is usually most undesirable!

17.4 Clock Skew

17.4.1 INTRODUCTION

Clock skew occurs when 'clock' signals to flip-flops or other clocked or strobed devices which should occur simultaneously are delayed either by transmission line effects or by differential gate delays so that the clocked elements receive their clock signals at different times. This is most serious in a circuit such as a shift register, where each flip-flop is supposed to set to the logical state of its neighbour prior to the clock signal. When there is no clock skew, such an arrangement gives very rapid shifts, but failures can occur if the difference in timing between clock signals to adjacent flip-flops exceeds the settling time of the devices. Then the flip-flop with the 'late' clock signal can be set to the output state of its neighbour after the latter has changed.

Clock skew can also be caused by the use of a very slow-edged clock pulse fed to devices which have different clock threshold voltage levels (see Section 11.3). In this chapter it is assumed that fast-edged clock pulses are used.

17.4.2 AVOIDING CLOCK SKEW

So far as is known to the author, there is no universally infallible cure for clock skew. Worst-case design must be used. Line delays must be calculated very carefully, and when there is a risk that clock skew might cause malfunctioning, an extra delay must be inserted in the signal line. If the clock skew is caused entirely by line delays, it can be eliminated in uni-directional shift registers or counters, etc. by positioning the device which drives the clock line at one end of the group of packages which comprise the register or counter, and routing the clock line such that the clock pulse 'flows' along the group of packages in the opposite direction to the data flow. That is, in a counter the clock line will be driven from the most-significant-digit end, or in a 'shift right only' register, it will be driven from the 'right-hand' (least-significant-digit) end.

There is no such 'easy' solution for the case of a register which must perform either left-shifts or right-shifts, or for a pair of registers which must interchange data on a single edge of the clock. Master-slave-type flip-flops can sometimes be used for the registers, so that all transfer of data is implemented in two stages, which eliminates all risks associated with clock skew. However, there can be other design considerations which dictate the use of single-edge-triggered devices such as D-type flip-flops. Then the problems of clock skew must be overcome. Most engineers tend to develop their own favourite solutions to such problems. The author's method with clock skew is to limit the number of devices on any clock line such that the line can be driven by an ordinary T.T.L. gate, then to position the gate which drives the line in the middle of the row of devices to be driven, so that each clock drive gate is driving two short lines in parallel. A shift register can be implemented in two adjacent rows of packages on a board, and two gates in the same package can be used to drive the two clock lines, again preferably from the centre of each line.

Clock line driving in this manner is against the advice offered in Section 18.2.2.3 to avoid allocating gates which will switch simultaneously to the same package, and, as explained in Section 16.1.2, driving the lines from the centre can cause steps. However, the half-lines should be quite short, and where clock skew is critical, slower edges which arrive virtually simultaneously represent the best practical compromise. The double value rail spike caused by the two clock-drive gates switching simultaneously has never caused any problem, presumably because the devices to which the clock signal is being fed have high rail to earth capacitances and they suppress the spikes more effectively than ordinary gate packages (see Section 16.2).

When more devices must be clocked than can be fed from a single package, then delays must be inserted in the signal lines between blocks of devices. A single Series 2 inverting gate can be used for the delay, with the output from the last stage in each block taken from the $\bar{Q}$ output instead of the Q so as to preserve the correct logical polarity.

The best solution to clock skew is being provided by the semiconductor manufacturers in the form of M.S.I. shift registers and integrated counters. The trend towards further integration may well make clock skew problems a thing of the past!

17.4.3 CLOCK DISTRIBUTION AROUND A SYSTEM

One almost certain way to guarantee trouble from clock skew is to 'daisy-chain' a single heavy current clock signal all round a system. On paper, with only one source of a clock signal, how can there be a problem? In practice, with taps all along a line, the delay down the line will be appreciable. It has been the author's experience that a low-current 'pyramid' distribution will give much better results, especially if all gates at any level in the pyramid can be allocated to the same package. When this is not possible, clock drive gates can be selected to have matched delays. In practice, most batches of T.T.L. appear to be quite consistent in delay, and the necessary matching is a simple task—the first dozen or so packages tested should yield an adequate match.

17.5 Current Hogging

17.5.1 STROBING OF LONG LINES

Long lines from one part of a system to another can be tolerated if the data on the lines is 'strobed' or 'gated' at the receiving end of the lines so that the data on the lines is 'sampled' only when the lines have settled out to a steady d.c. condition. However, there is one risk in such strobing of data from long lines. A line may be in the 'low' state, but not carrying any current from the strobe gate at its end because the device which controls the strobe gates is also low, and is 'hogging' all the base current from the input transistor in the strobe gate. When the control gate goes 'high' to enable the strobe gate, the base current must establish itself in the long line, and a pulse will be seen on the input to the strobe gate. The width of the pulse is a function of the length of the line.

When a single strobe gate terminates a 100 Ω line, the amplitude of the pulse should be safely below the switching threshold of the gate, but if the line feeds a fan-out of more than one, or if the line is a higher impedance line, then the disturbance could become troublesome.

Such a disturbance on a long line can be avoided by ensuring that the control gate can sink current only from those strobe gates whose data lines are at a high level, and that all 'low' data lines carry the current from their strobe gates. The simplest way to ensure this is to insert a germanium diode in series with the output of the control gate, so that the effective low voltage output of the gate is raised 0·2 to 0·3 V above that of the gates which drive the lines.

In high reliability equipments the use of germanium devices might be prohibited. In such cases, a resistor or a silicon power diode can be used, but only if the track from the control gate to the strobe gates can be guaranteed to be virtually noise-free, and the diode used needs to be carefully selected to have a low forward drop at currents up to 20 mA.

17.5.2 MATRIX DRIVING

In situations such as matrix driving, cases can arise when it is known that a gate which is driving a high fan-out will not in fact be called on to sink current from all the gates to which it is connected because some of the gates will have other inputs connected to other gates which will also be sinking current. In such cases only one gate can sink all the input current available from any gate in the matrix, and the rated '0' level fan-outs can be exceeded. ('1' level fan-outs can be exceeded only if the designer assumes that few, if any, of the driven gates will have leakage currents which approach the specified limit values.) In these cases no precautions need be taken to ensure uniform current distribution. A drive gate with an exceptionally good output transistor may in fact 'hog' all the current from its row in the matrix, but it will be able to do this only if it presents a lower impedance to the matrix than all the other ON gates, so provided that all gates work within their specified limits, no harm can result from this form of current hogging.

17.6 External Signals to T.T.L. Circuits

17.6.1 SLOW-EDGED PULSES

The output of a T.T.L. device changes as the voltage on its input is swept through the active switching threshold region, which is normally less than 100 mV wide. When a T.T.L. device is driven by another T.T.L. device, the change of voltage is fast, and the input is swept through the active region before the output of the device can start to change. However, if a very slow input signal is applied, the output can start to change while the input is still within the active region. Under these conditions the T.T.L. gate forms a high-gain amplifier, and any feedback can cause oscillations. Feedback can occur via the supply rail to the device involved, via the connections on the printed circuit board on which the device is mounted, and via the parasitic capacitances on the chip itself. Thus if the input to a device is held in the active region for a period longer than the delay of the gate it is probable that oscillations will occur.

The device specifications define an absolute lower limit to the active zone ($V_{IL\ Max}$) and an absolute limit to the upper end of the zone ($V_{IH\ Min}$), but the actual active region on any device is very much narrower than the difference between these limits. A typical allowance of 100 mV can be made, which, with a typical Series 1 gate speed of 10 ns, allows edge speeds of down to 10 V/μs, or about a quarter of a microsecond for a normal '0' to '1' level transition. For a Series 2 device, 100 ns can be allowed for a normal level transition. If gates are to be controlled by pulses with edges slower than these limits, the edges should be 'sharpened' by some form of feedback circuit such as a Schmitt trigger circuit.

17.6.2 SWITCHES

When switches or relay contacts are used to control logic circuits, there is always a risk that 'contact bounce' can cause spurious switching. This can be demonstrated very simply by connecting the switch to the 'clock' input of a counter circuit and observing the count for each operation of the switch. Sometimes the count will change by only the desired 'one', but more usually it will count between two and five input pulses for each operation.

Such 'contact bounce' can be overcome by connecting the switch to a 'back-up' bistable. The switch should form a changeover, with the centre contact earthed and the two outer contacts taken to the 'R' and 'S' inputs of the bistable. Thus one input to the bistable must be 'high' (open circuit) before the other input can be earthed, and the first application of the earth to that input will change the state of the bistable. A subsequent break in the earth line will not disturb the setting of the bistable—the earth must be deliberately applied to its other input before it can change its state. If the switch is in a 'noisy' environment, a two-pole changeover switch can be used with the second pole connecting the 'high' input to the 5 V supply rail via a 1 kΩ resistor. The bistable should be situated adjacent to the logic it controls so that any long wires or tracks will be on the inputs to the bistable.

Similarly, if a system diagram calls for a signal line to be 'opened' by a switch, the 'opening' of the line should be done by a logic gate and the second input to the gate should be controlled by the switch with a suitable 'back-up' bistable.

In some cases it may be desirable to route data lines via a rotary selector switch which will be pre-positioned before the circuit is switched on. Such use of a rotary switch cannot cause any failures because of contact bounce, but failures can be caused by cross-talk in the wiring to the switch, and unless the switch can be mounted very close indeed to the circuits it is to control, it is usually better to use integrated circuits to perform the actual data routing, even though this may involve the use of a whole row of back-up bistables.

18

The Influence of T.T.L. on System Design

18.1 Design of the Logic System

In the early days of digital logic systems when either valves or discrete transistors were used as the driving elements and the logical decisions were performed by 'wired-OR's and resistor or diode networks, any digital system could be designed (logically) by a mathematician who did not know how the equipment was to be built. 'AND's and 'OR's could be realized in whatever configurations were needed, and if loading rules for one type of transistor were broken there were other types which could be used. In fact, the logic elements could almost be tailor-made to suit the system design.

To a much lesser extent this is also true of the D.T.L. integrated circuits where the 'wired-OR' can still be used, and where discrete diode 'extenders' can increase the fan-in of a device. With T.T.L. quite the reverse applies. The 'wired-OR' is no longer generally available, and both fan-in and fan-out are restricted. Instead of the gates being built up to fit the system, the system must be designed to suit the range of devices available.

This means that if good designs are to be achieved, every person concerned with the design needs to be fully conversant with the range of devices available and also with the rules which govern their interconnection and use.

Right from the time a project is first conceived, decisions have to be taken which have their effect on the detailed design of the final equipment —especially on the design of the printed circuit boards.

The knowledge that the largest 'AND-OR-INVERT' function available without expansion is the 'quad-two' should encourage system designers to group the outputs of registers or other elements which require a gated OR function into blocks of four or fewer. Similarly, the sensing of 'all ones' is simple if done in blocks of no more than eight lines (when a single eight input NAND gate can be used) but presents real problems on groups of nine or ten lines. Also, it may occur that in the early stages of the system design, sensing on a group of lines could be either for 'all ones' or for 'all zeros'. If the system designer knows his logic families, he will opt for the 'all ones'. Sensing of 'all zeros' requires either the use of an inverter from every line to yield 'all ones', or of positive exclusive-OR gates—which are available only in a quad-two package, and are expensive, with a higher dissipation than simple NAND gates. To the theoretical system designer who never sees the practical hardware which implements his ideas, these points may seem of very little significance. To the engineer who has to struggle to build the equipment down to a cost, and has to face the problems of packaging and cooling, such points as these can mean the difference between success and failure.

18.1.1 THE INFLUENCE OF M.S.I. ON SYSTEM DESIGN

A glance at any integrated circuit manufacturer's current list of M.S.I. devices will show that there are some very useful devices available, but if the true savings which they make possible are to be realized, this availability must be known in the early stages of the system design. It can happen that decisions taken early on in the design of a system can preclude the optimum use of readily available standard items.

If, as often happens, the system is sold to a customer before the detailed design is undertaken, some decision taken in all good faith but without full knowledge of 'state of the art' integration can leave a design engineer with the problem of having to design some special logic circuitry to modify the working of an M.S.I. package, or, even worse, of having to find a manufacturer who is willing to make special 'custom designed' M.S.I. circuits at a price which is not totally prohibitive. On the other hand, had all the facts been appreciated early enough in the design, the customer's requirement might have been met perfectly well by a different specification which would have allowed the design engineer to use the standard M.S.I. package without any peripheral gating to modify its function.

18.2 Implementation of the System

18.2.1 SUB-SYSTEMS

The practical implementation of a logical system usually involves splitting the system into sections to form sub-systems, which are in turn split into a number of circuits which can be fitted on printed circuit boards.

If the sub-systems are to be separated by any distance more than a few inches, they will generally be interconnected by properly terminated balanced highways, and in systems where extra sub-systems can be added as required, special tapped highways will usually be provided. The design of such highways is not simple, and their realization may be expensive, particularly if special cable has to be used. Thus it is usual to take great care in splitting a system into sub-systems to ensure that the interconnections are minimized and that the various interfaces are correctly specified.

18.2.2 DIVISION OF THE SYSTEM INTO PRINTED CIRCUIT BOARDS

Although the division of a system into sub-systems is usually done very carefully, division of the sub-systems (or of a whole, single unit system) into boards is sometimes not done quite so well.

18.2.2.1 *Board size*

The size of the printed circuit board to be used needs to be carefully chosen. In some instances, especially in airborne equipment, the units might have to fit into standard cases, and the design team have virtually no choice in the question of board size. When no such restrictions apply, the board size should be chosen as an essential part of the main logic design process, and it should not be left to a mechanical design team or to the drawing office to select dimensions for the board.

Mechanical considerations must not be overlooked when the board size is selected, but efficient electrical or logical design should have priority. In most systems, the printed circuit boards are the most expensive single components used, and often the total cost of the boards involved will be about the same as the total cost of the integrated circuits which they carry, so it is well worth while putting some care and effort into the design of the boards!

Many of the factors involved in the choice of board sizes are discussed in *Printed Circuit Boards for Microelectronics*, Chapter 9, but if the best possible results are to be obtained from T.T.L., there are some other factors which need to be considered. The possibility of using the board dimensions as a 'semi-automatic' limit to track lengths is mentioned in Section 16.1.5. However, this should not be regarded as a primary means of selecting board size, but rather as a handy bonus if it can be applied after the other electrical and logical points have been considered.

18.2.2.2 *Functional division*

The first consideration should be the satisfactory division of the system into boards in such a way that each board can carry a complete working function or sub-function. Such a division reduces the number of board to board connections, and it also enables boards to be re-designed at a later stage (after the equipment has been in service) if (say) developments in M.S.I. offer a substantially better way of performing the function. A complete functional split will also reduce test costs. The cost of testing a board which carries the whole of a particular logical function may be greater than the cost of testing any of (say) three smaller boards into which the function might be divided, but the cost of testing the entire function will be lowest if only the one large board has to be tested.

Most large digital systems have to handle more than one 'bit' of data. Commonly used sizes for systems are 8 bits, 16, 24, 36, or 48 bits. Systems usually contain a multiplicity of registers, arithmetic units, comparators, input/output stages, etc. When a system is being split into boards, it is tempting to take each of the large 'boxes' on a block diagram and make it all on one board. However, this is often not the best approach when T.T.L. is to be used. The overall system block diagram may well include several 'OR' functions in its links between blocks, which could mean that the output from one board could have to be fed into another board simply to 'OR' with the output of the latter board, the output from the 'ORed' function then being taken to a third board. In Fig. 18.1, which shows a hypothetical data handling sub-system, the 'OR' in the right-hand data link is an example of how such wiring can become necessary.

Another major disadvantage of such a split into separate individual functions is that in a parallel working system, a substantial number of devices on a board would have to switch simultaneously, and cross-talk and rail spike problems could become troublesome in a large system.

If the system contains more than 8 bits, it is probable that the control signals to the various registers, etc. will have to be 'buffered' or amplified by gates on the register boards (etc.). When this is done, it is often possible to perform some of the control logic on the data boards, especially the selection of timing strobes or clock lines to the registers. A large system will require a multiplicity of such 'buffering' gates for each logical stage, and each 'box' on the block diagram will thus contain a number of identical sub-functions (see Fig. 18.2).

Much better working in T.T.L. systems will be achieved if these sub-functions are separated, and each board contains a complete 'slice' of the entire data handling sub-system. If the sub-system shown is split into six 8 bit boards as shown in Fig. 18.3, the board size will remain the same, but all boards in the sub system will be identical, so spares holding will be cheaper; only one sixth of the devices on any board will be switching simultaneously (assuming that all devices in a sub-function switch together); and 19 data connections per bit to the back wiring will be saved. Also, fault diagnosis to a board will be much simpler. If the sub-system is built with a board for each of the 'boxes' on the diagram as in Fig. 18.1, a faulty board could be found only by examining the signals on the inter-board links in the back wiring. With the sub-system built on six identical boards as in Fig. 18.3, a fault on any one board will cause only the outputs from that board to fail, and it will be obvious which board is faulty. If all

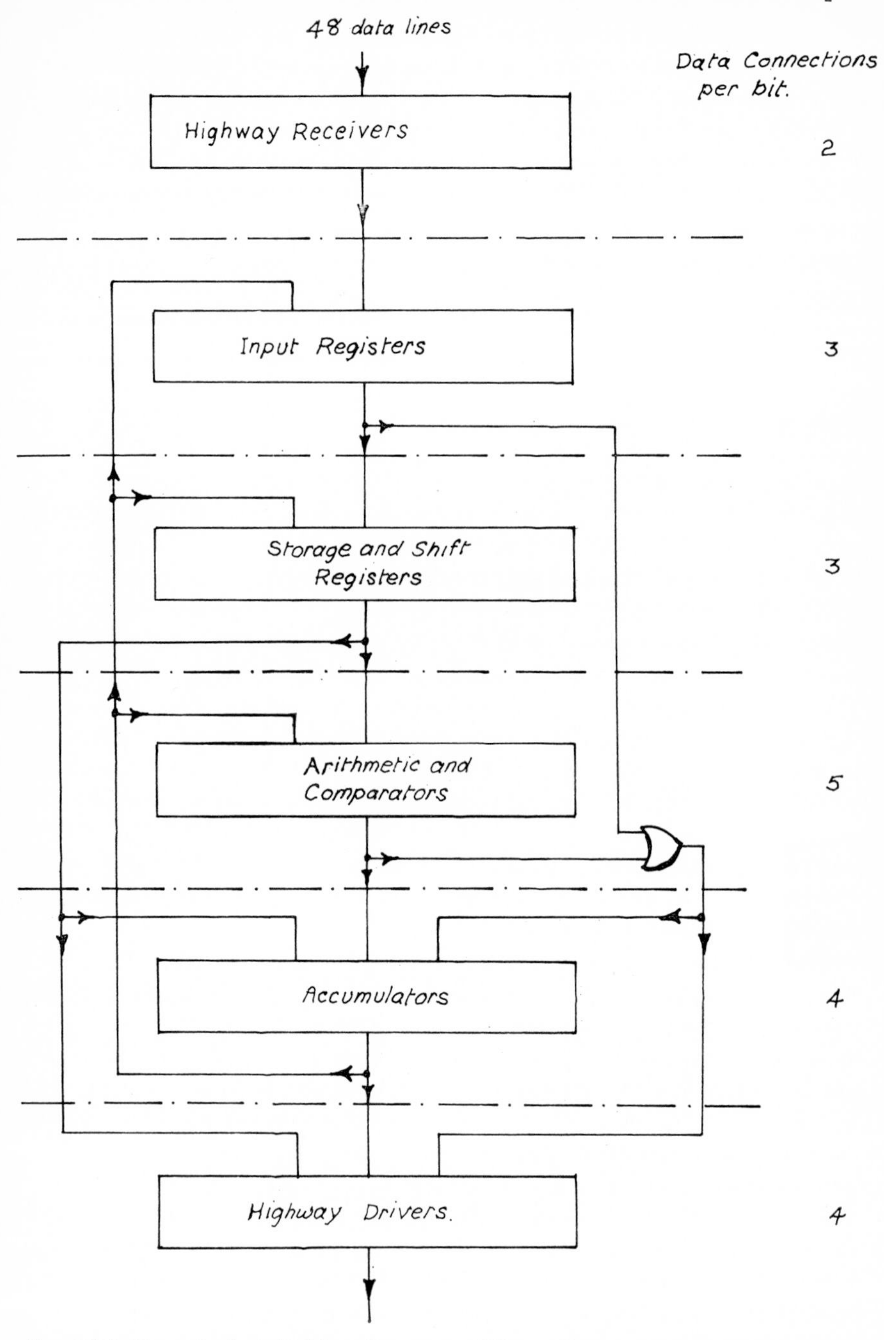

Fig. 18.1 Hypothetical 48-bit sub-system. (Note—Data connections are shown for one bit only. Control connections not shown.)

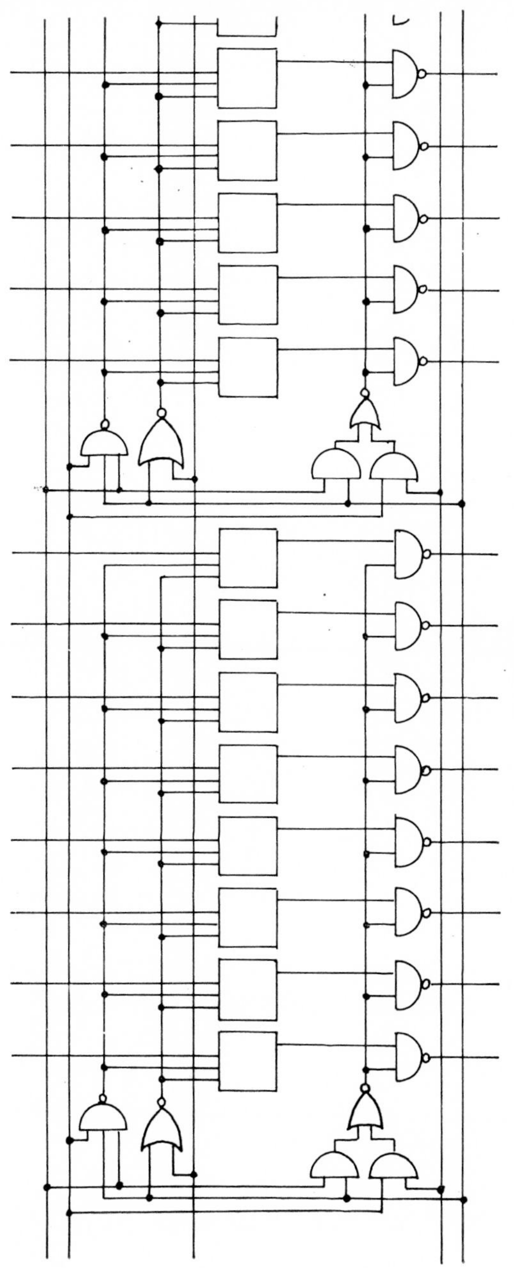

Fig. 18.2 Multiplicity of control line gates in portion of large register.

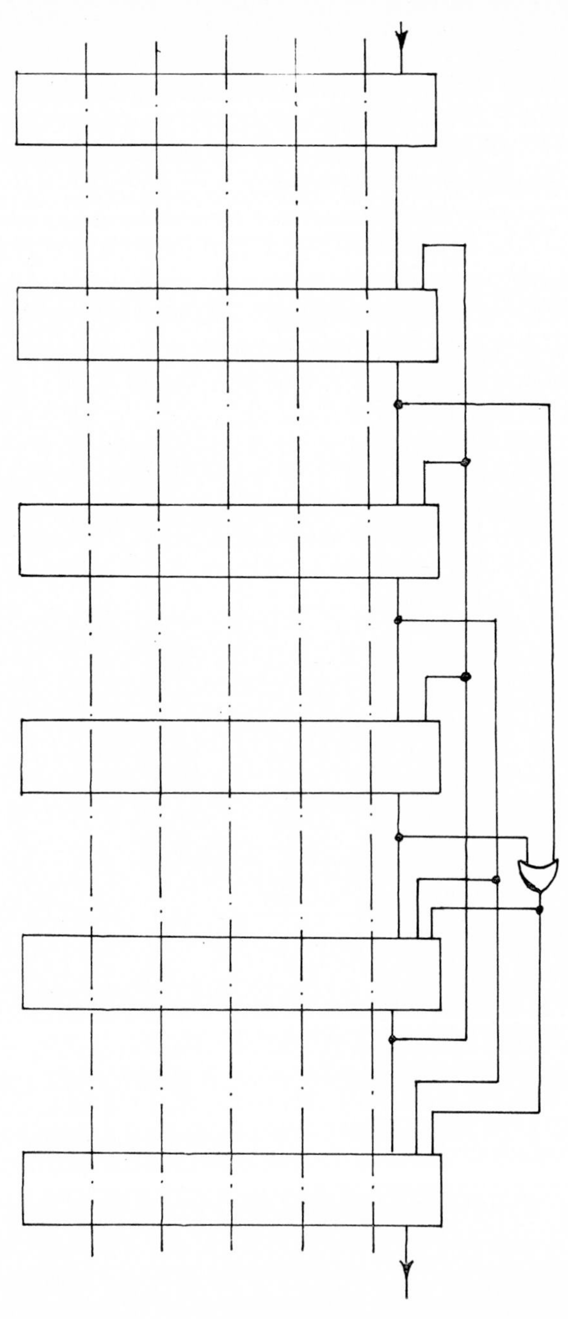

Fig. 18.3 Functional split into identical boards. (Note—Data connections are shown for one bit only. Control connections not shown.)

48 bits of the output fail, then the fault must lie in the control area, which will be common to all six boards (not shown on the block diagram). Finally, the dissipation on all six boards will be equal, which will simplify thermal design slightly.

18.2.2.3 *Allocation of gates to packages*

A similar principle on a smaller scale should be applied when gates on a logic diagram are allocated to packages. Whenever possible, multi-gate packages should contain gates which are cascaded in the logic diagram, rather than gates which work in parallel. In the case of registers which are implemented with dual flip-flop packages, it is better to include pairs of bits which have a common data input in one package rather than to include pairs of bits with common clock or control signals. Fault diagnosis to a package will be simplified if multi-gate packages contain gates which are connected directly in cascade. If one of two such cascaded gates fails, the output from the package will be false, and it is not necessary to test to see which of the two gates has failed, whereas if the two gates were in separate packages, an extra test would have to be made to determine in which package the fault lay.

19

Verification of Logic Design

As the size of the logical functions in individual packages grows, proper verification of the logical design of a system or a board becomes increasingly important. In the days of discrete component circuits, many design errors could be rectified by the addition of a diode or two and some simple wiring changes. With M.S.I. this is not possible, and a small slip in design, or in the understanding of how a particular M.S.I. element functions, can mean that an entire printed circuit board has to be completely re-designed.

Work is proceeding on computer programs to check-out entire logic designs (and to design the printed circuit boards), but even when such programs are in full use, there will probably always be circumstances in which a design must be checked out manually.

A full check is not difficult, but it must be done methodically. It is quite useless to 'have a good look' at a circuit, or to check one or two parts of a circuit which are known to be critical. If the designer has worked hard on a critical part of a circuit, it is unlikely that there will be an error in his final design. Errors occur mainly in the 'simple' parts of the circuits. The classic errors are a bistable which can set to either logical state during 'run up', the omission of a strobe gate, and logic races.

19.1 Interface Specifications

Omission of a strobe gate can occur because, when a system is split down to board level circuits, the interfaces are inadequately defined. A simple example of this type of error can occur when 'shift-left' or 'shift-right' functions are used. The control portion of the system may contain an 'overflow detector' to detect and hold a '1' which is shifted out of the working range of the machine. A simple method of making the connection for such an overflow detector is to connect to the output of the end bit of the register (etc.) which drives into the shift stage. (This is a point which might well be brought out to the back wiring anyway.) If the data-handling portion of the machine is designed by one engineer and the control section by another, each might assume that the other has included a timing gate in the overflow line. If neither of them includes such a gate, an overflow will be detected every time the end bit of the register handles a '1', whether the shift function is to be implemented or not.

There is only one cure for this type of error, and that is the provision of a full interface specification. Any engineer who is asked to design a part of a system, or a single board, should demand, and should be given, a full specification which defines clearly all the inputs to his part of the system, and specifies all the outputs he must provide. If the interface is between standard T.T.L. devices, it is not necessary to define voltage levels or edge speeds, but the logical states must be clearly indicated, and if '1's and '0's are used, the specification must define the logical convention used (i.e. whether '1' represents a 'low' voltage of about 0·3 V or a 'high' voltage of about 3·5 V). It is also essential to state the fan-out, in terms of T.T.L. gates, which each line can drive, and to define the timings of all input pulses.

19.2 Circuit Checking

Logic circuits can be checked absolutely only by defining the state of all devices in a circuit at any instant in time. This is not so difficult as it may sound. If no computer program is available, the easiest way to begin is to number every device on the circuit, and to number or letter all the input connections to the circuit.

19.2.1 NUMBERING OF DRAWINGS

The device numbering should be done in such a way that every gate or function in a package which has inputs which cannot be interchanged with the inputs to another gate in the same package has a unique number (see Fig. 19.1). For instance, in a J.K. flip-flop, the J input gate, K input gate, clock line, direct set, and direct clear inputs would all have separate (consecutive) numbers. Any device inputs which have a higher loading than that of a normal T.T.L. input should be given some extra identification so that they can always be recognized as 'specials' (e.g., add 1000 to the number the input would normally have had).

NAND gates have a single number regardless of the number of inputs to the gate (so that a 'quad-two' package will contain four numbers, which need not be consecutive), and AND-OR-INVERT gates have a number for each input AND gate but the internal OR function is un-numbered. On all complex functions which use more than one number, the numbers should run consecutively whether or not the parts of the function are shown adjacent on the logic diagram. Otherwise numbering should proceed from left to right or up and down over the drawing, or in some other logical manner such that any number can be found in the minimum of time. Such

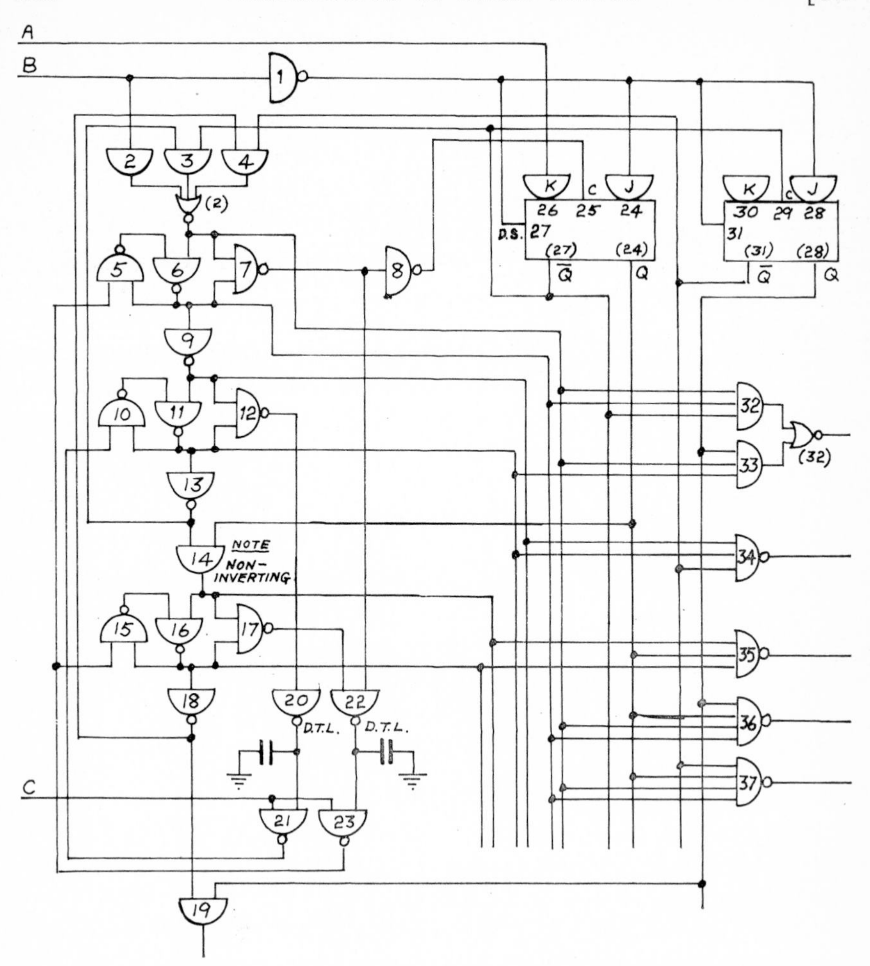

Fig. 19.1 Logic diagram showing numbering of elements. (Note—This is a simplification of a much larger circuit. A number of gates and inputs to gates have been omitted to leave a 'mimimized working circuit'.)

numbering should have no reference to the ultimate position of components on a printed circuit board.

19.2.2 INTERCONNECTION SCHEDULE

At this stage it is helpful, although not essential, to compile an interconnection schedule. One possible form for this is to list in columns the device numbers, device types, points from which each input to the device is fed, and the devices fed by the device under consideration (see Table 19.1).

TABLE 19.1 Schedule of Interconnections

Gate or component	Type	Inputs fed from	Outputs feed to					
A	Input	—	26					
B	Input	—	1	2				
C	Input	—	21	23				
1	610	B	24	27	28	31		
2		B — —	6	7	32	33	36	37
3	133	13 27 —	—					
4		18 31 —	—					
5	420	6 23	6					
6	420	2 5	5	7	9	32	36	37
7	420	2 6	8	22				
8	610	7	25					
9	610	6	11	12	34			
10	420	11 21	11					
11	420	9 10	10	12	13	33	34	
12	420	9 11	20					
13	610	11	3	14				
14	420 N	13 24	16	17	35			
15	420	16 23	16					
16	420	14 15	15	17	18	35		
17	420	14 16	22					
18	610	16	4	19				
19	420 N	18 28	Ou	tp	ut			
20	610	12	21	C1				
21	420	C 20	10					
22	420	7 17	23	C2				
23	420	C 22	5	15				
24	J	1	14	35	36	37		
25	F/F C	8	—					
26	K	A	—					
27	DS.	1	3	29	32			
28	J	1	19	33	36			
29	F/F C	27	—					
30	K	—	—					
31	DS.	1	4	34	37			
32	142	2 6 27 —	Ou	tp	ut			
33		2 11 28 —						
34	330	9 11 31	Ou	tp	ut			
35	330	14 16 24	Ou	tp	ut			
36	240	2 6 24 28	Ou	tp	ut			
37	240	2 6 24 31	Ou	tp	ut			

Outputs of complex functions are listed under the lowest number (or numbers) of their input gates, and inverted ($\bar{Q}$) outputs can take the highest number in the input set.

One benefit of such a list is that it can automatically indicate whether any fan-out rules have been broken. *Pro-forma*s can be drawn for these lists, and the last column can be lined to form small boxes, the number of boxes for each line of the list to be equal to the fan-out of the logic family to be used. Any driven gate number in the '1000' series is written into the list over two (or more) boxes, but all standard inputs are filled in one number to each box. Thus any excessive fan-out will be obvious because there will be a number which has no box available. Buffers or other devices which have a higher than normal output capability occupy three lines of the list so that their higher fan-out is accommodated.

Discrete components can also be accommodated on such a list. Resistors and diodes are regarded as single input, single output devices, and transistors are regarded as having two non-interchangeable inputs, the base having the higher number than the emitter, and one output, the collector, which takes the same number as the emitter. 'Fan-out' for such discrete components cannot be checked automatically by the 'boxes' described above, and it is better to use a separate *pro forma* for discrete devices.

19.2.3 LOGIC CHECKING

When the drawing has been fully numbered and an interconnection list has been compiled, checking can be started. This is done by taking a large sheet of graph paper (inch by tenths is ideal). The numbers for all gates are listed in alternate squares down the left-hand side of the sheet, together with the symbols used for the inputs to the circuit. The logical 'rest' state of all inputs to the circuit is then indicated as 'low' or 'high' by a line at the bottom or top of the appropriate square in a row to the right of the column of numbers.

Each 'low' input thus marked which goes to a NAND gate must set the output of this gate 'high' regardless of the state of the other inputs to the gate, so these can be put in, and any NAND gates with all inputs 'high' must have 'low' outputs. Other types of gates are considered in turn, and the 'rest' condition of the circuit can be determined (see Fig. 19.2). If the circuit contains bistables or 'clocked' devices, it may be impossible to determine a rest state for them. Their output lines have to be left blank until an input signal 'sets' or 'clears' them.

When the rest state has been determined (and it has been checked that this is as intended!), a vertical line to the right of the horizontal lines which define the rest state is chosen and marked as 't_0', and the first input to the board or circuit is drawn as changing its state on this line.

The change of state of an input to the board should cause one or more devices on the board to change state. These changes of state are all indicated by 'L's for devices which turn on, and inverted 'L's for devices which turn off. The vertical lines of the 'L's are drawn one square to the right of line t_0. The 'L's should *not* be joined back to the 'rest state' horizontal lines.

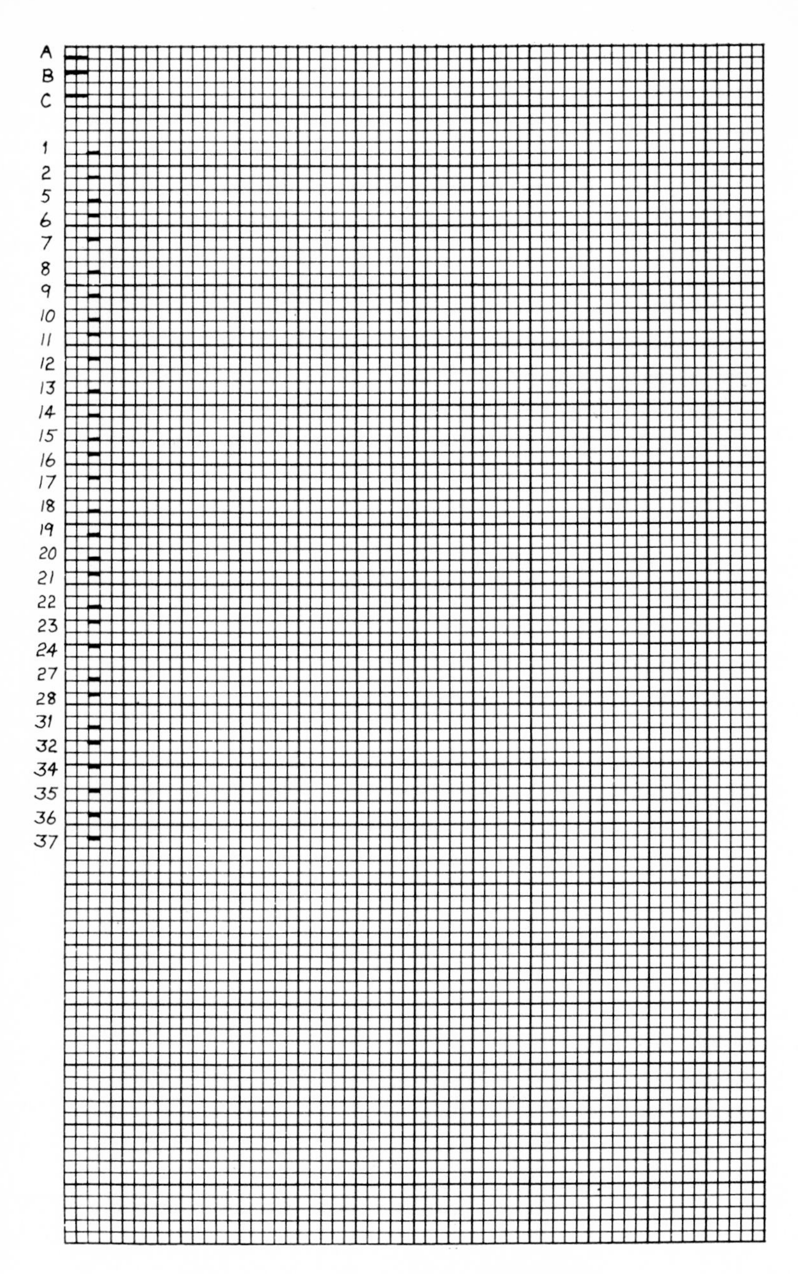

Fig. 19.2 Initial conditions determined.

When all the devices affected by the input signal have been dealt with, the output of the first device which changed can be considered. From the interconnection schedule or the drawing, all gates fed by this device are considered in turn, and the appropriate 'L's are drawn in as necessary, a further square to the right.

When the entire fan-out of the first device has been considered and all the changes it causes have been marked, the horizontal line which joins the 'L' for the first device back to the rest state line can be drawn. This procedure is then repeated for all devices until a stable state is reached, or until it is obvious that the circuit is re-cycling steadily.

The key to the whole system of checking lies in not joining the 'L's indicating a change of state back to the rest state line or the last change of state until after the entire fan-out driven by a device has been considered. The work can be left at any time, and it is always obvious how far the check has proceeded (see. Fig. 19.3).

Sometimes it may not be obvious from a logic diagram which signal has caused a particular gate to change, especially if several unrelated changes occur (nominally) simultaneously. Any possible confusions on the timing diagram can be resolved by noting against an edge on the diagram the number of the gate (or other device) which caused the change. For example, in the circuit shown in Fig. 19.1, gate 23 going low causes gates 5 and 15 to go high. The number '23' could be written on the diagram (Fig. 19.3) before the inverted 'L's for gates 5 and 15.

Because an output of a complex device or an AND-OR-INVERT gate carries only one number, it may not be obvious which of its input gates has caused the output to change. This can be clarified by noting the number of the input concerned against the step on the diagram (see 2 and 32 on Fig. 19.4).

When a steady state (or a cyclic state) has been reached, a gap can be left, and a new vertical line can be chosen for the next input signal to change. A circuit may reach a steady state when it generates an output signal, and that output signal may cause other parts of the system to alter the input signals to the circuit under consideration after a pre-determined time delay. Only sufficient space need be left on the diagram to make it clear that a delay has occurred. Delays, whether external or internal to the circuit being checked, can be indicated on the diagram by an arrow and a letter D to define the square (or squares) in which the delay occurs.

The resultant picture shows clearly the logical state of every device at any time, with a time scale of one gate delay per square. It assumes that all gate delays are substantially equal.

On most circuits, the condition of a particular input signal may affect only a small portion of the circuit. In such cases, only a partial timing diagram need be drawn to verify the working of the sub-circuit involved

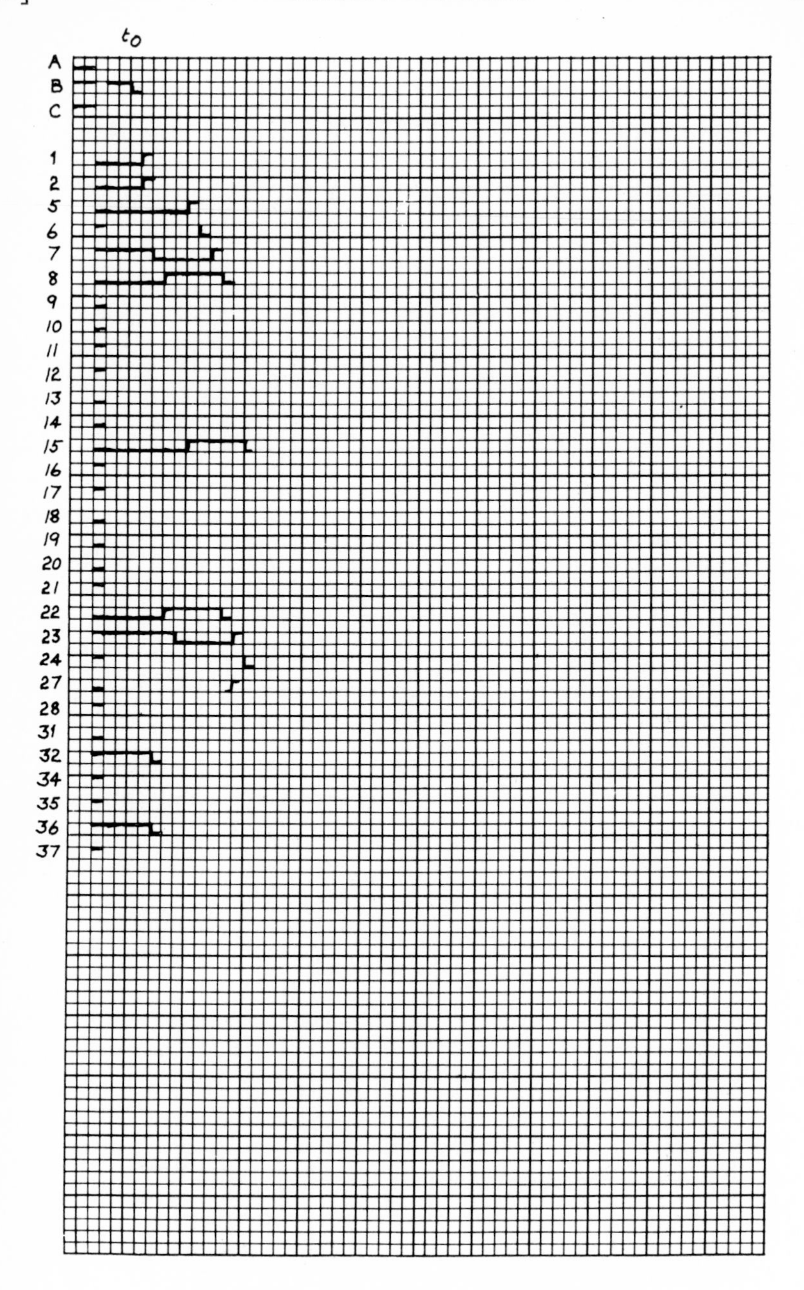

Fig. 19.3 Checking stopped as Gate 6 changes.

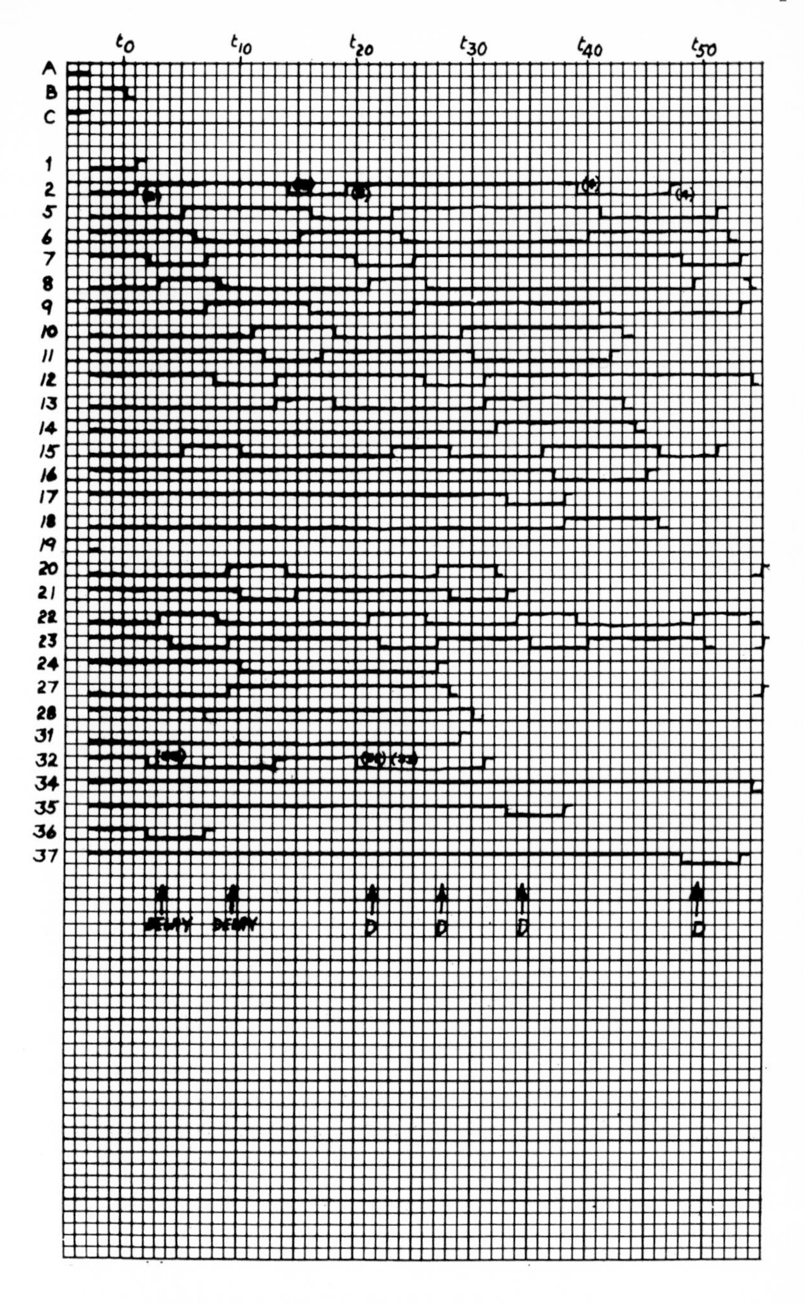

Fig. 19.4 Check completed over first 55 steps.

when this particular input signal changes from the state which was assumed for it in the complete timing diagram.

19.2.4 VARIABLE DATA INPUTS

Such a system can give an 'absolute' result on a circuit such as a timing or control circuit where the states of all input signals are known, but on data handling circuits, the data (which may be 24, 48, or more bits) may be '0' or '1', and there will be rather too many combinations to draw them all out in full! The easiest method of checking such circuits is to call the data lines D1, D2, D3, etc., and when an inversion of data occurs, to draw the vertical line to indicate the timing, and to write 'f D1' or 'f $\overline{D1}$' after it. When a data line is gated such that its condition is known while the gate is inhibited, the line indicating 'low' or 'high' state is drawn in the usual manner.

Combinational logic, such as a read-only memory, or other devices with a complex truth table, can be treated in a similar manner.

19.3 Errors in Logic

19.3.1 LOGICAL ERRORS

When checking is properly carried out, it may happen that something does not work out as it was hoped it would. A bistable may be 'set' when it had been assumed that it would be in the 'clear' state, or it may be found that an inhibit signal to a gate does not get there until after the pulse it was intended to suppress. When (not if !) such design errors are found, the full working on the diagram makes their rectification relatively simple. The diagram will show which gates in the circuit can be used to provide a signal of the appropriate logic state to perform the desired function. Usually, one has only to look for any line which 'goes low before this time, and high again after that time' (etc.). The limits 'this' and 'that' should be set as wide as possible. A signal from quite a different part of a circuit can often be used to provide the necessary function. Odd links put in in this manner can make the logic drawing look rather hard to understand, but provided a concise description is included in the handbook for the unit, it is better to end up with a cheap circuit and a complex logic diagram than with a simple drawing and a more expensive product!

As alterations are made to the logic drawing, the timing diagram can be corrected. Unless it is absolutely certain that the alterations have corrected all errors, it is advisable to mark the corrections in coloured crayon on a Xerox copy of the original timing diagram. The original can be finally corrected when the whole design has been proved correct.

19.3.2 TIMING ERRORS

A basic logic check of this nature, where one square on the graph paper represents one gate delay, will indicate any gross errors in timing, but it will not, as drawn, show up what could be dangerous race conditions as faults. Races occur whenever two (or more) inputs to a device have followed different paths through the logic so that their relative timings may be uncertain. A typical condition is that one input to a NAND gate is low, and one high. The high input goes low, and then the low input goes high. The output of the gate will supposedly remain high throughout. However, if all the gates in the line feeding the input which supposedly goes low first are on or near their slow limit, and all the gates in the other input line are fast, then the rising input can change first, so the NAND gate gives a spurious negative pulse output. Similarly, a 'wanted' pulse may not be present because a gate is turned OFF by one signal before it has been turned ON by another.

The timing diagram must be studied to see whether any such race conditions are likely to occur. As a very rough 'rule of thumb' guide, circuits are likely to be trouble-free if the 'late' signal passes through more than three times as many gates as the 'early' signal.

Worst-case tolerancing of races is virtually impossible, as only 9000 Series T.T.L. has a minimum tolerance on all gate delays, and that tolerance demands a ratio of over 5 to 1 between the 'slow' and 'fast' lines to guarantee trouble-free operation.

Races can be eliminated by using the 'early' signal to enable a gate in the 'late' signal line. When this is impracticable, a cure for a race condition can be provided by using an open collector gate (or a D.T.L. gate) in the 'late' signal line, and allowing provision for a capacitor to be fitted on the output from this gate when the board is tested. When this is done, the delay must be arranged to occur on a rising edge. Such capacitors can be fitted in 'worst-case-toleranced' circuits, and the value of the capacitor can be calculated when the circuit is designed. Such worst-case working cannot be recommended, though, as a typical result was a circuit which had to produce two successive output pulses, each with a guaranteed minimum width of 100 ns, on which the calculated maximum pulse width came out well over a microsecond! If capacitors are to be fitted, it is much better to 'select on test' and fit a capacitor which yields the desired result (with due allowance for drift caused by ageing).

Logical verification may seem tedious, but it is well worth doing. A 150 gate circuit can be checked through over 300 gate delay operations in less than 3 days work, and minor corrections to the circuit (found as a result of the check) can take a fourth day. Against this 4 days 'lost' must be balanced the cost of re-making the artwork and the board, and all consequential delays, as well as the time spent finding the error, when the original design is found to be faulty when the first board arrives in test or commissioning.

Generally, the design of the board will have to be checked out sometime, and it will always be cheapest to do this once, properly, at the earliest possible stage. The timing diagram described here provides an all time record that the test has been carried out, and it can often be used in a handbook, etc. to help to explain the working of the unit.

Computer programs for logic checking can produce a comparable result to a manual check in a fraction of the time, but even if such a program is available, there can still be circumstances in which a manual check will have to be done. The translation of the logic circuit into the necessary input form for the computer can take nearly as long as a manual check, and if a large system is being designed in a short time scale, the computer facility may become overloaded.

If worst-case timings are to be calculated throughout, the use of a computer is probably the only economic solution for all but very small systems. Timings may be calculated for a manually prepared diagram, but this takes a long time, and great care has to be taken to ensure that the correct figures are taken for the start of each calculation. It is very easy to base one calculation on the assumption that a particular gate is on its slow limit, and another calculation on the assumption that the same gate is on its fast limit. This is probably quite correct working, but errors occur if the results of both the calculations are later applied to another part of the circuit! In circuits of any complexity, it might be necessary to record as many as a dozen different possible timings for a given edge, and extreme care has to be taken to ensure that the right value is used for subsequent calculations.

Appendix I

Guidelines for the best use of T.T.L.

The following points are offered as a quick guide to the best economical use of T.T.L. In almost all cases there is no sharply defined limit beyond which a circuit will not work, so the points offered can hardly be called 'rules'.

They may be regarded as rules, but in most large systems it is probable that one or more 'rules' will have to be broken if a complete economic design is to be achieved.

How many rules are broken, and by how much, must be left to the engineering skill of the designer. Wherever it is likely to be of any help, a reference back to the sections in which the parameters concerned are discussed is given in parentheses.

No apologies are offered for Guidelines 1–4 (which should be regarded as absolutely unbreakable rules). Failure to observe these 'obvious' points has probably led to more wasted time and money than any other single design fault or error.

General

1. Read the manufacturer's data sheet—including the fine print.
2. Do not exceed the limits quoted in the manufacturer's data sheets (see especially flip-flops—11.3).
3. Design the system around the logic functions which are known to be available—and make sure they *are* available at a reasonable price and not just 'advance information' on a device that may never be made. (18)
4. Check the design before the boards are made and the system is built—not after. (19)

System and Board Design

5. Aim for 'serial' or 'cascaded' working on a board rather than parallel working. (18.2.2.2)
6. Aim to keep the origins of all ORed functions—together with the package in which the OR is performed—on the same board. (18.2.2.2)
7. Minimize the number of board to board interconnections. (18.2.2.2)
8. Aim to locate gates which must switch simultaneously in separate packages. (16.2.4, 16.3.3)
9. Ensure that each board has a complete earth and power distribution grid. (16.2)

10. When packages are allocated to board positions, aim to put devices which will switch simultaneously on different 'legs' of the power distribution grid. If they must be on the same 'leg', aim to space them such that they can be separated by devices which will switch at other times. (16.2.4.3, 16.3.3)
11. Keep expander packages adjacent to the gates they expand and keep the interconnections short. (7.4.1)

Gate Connections

12. Earth one input to each unused NAND gate or each unused AND gate in an AND-OR-INVERT package or complex function. (8.2.2, 9.5.4)
13. Connect 'unrequired' inputs to a working input on the same gate whenever this can be done without breaking the fan-out rules for the driving gate. Such connections can sometimes be made even if the '1' level fan-out would be exceeded. (9.2.5, 9.3.2)
14. If gate inputs must be held 'high', they may be connected to the 5 V supply rail via a 1 kΩ resistor. One such resistor can be used for up to fifteen inputs. (9.2.5)

Tracks

15. Aim to use tracks of such a width that the characteristic impedance will be about 100 Ω or over. (16.1.2) Tracks 0·015–0·020 in wide on one sixteenth of an inch thick double-sided G10 material are suitable.
16. Aim for a minimum clearance between tracks of 0·030 in. (16.1.3)
17. Aim to keep all tracks less than 12 in long. (16.1.4)
18. If no precautions are taken to minimize cross-talk (K_C) in the wiring design, tracks should not run adjacent for more than 9 in for Series 1 T.T.L. or 2·3 in for Series 2 T.T.L. (16.1.3)

Thermal Design

19. Never allow any chip temperature to exceed 220°C. (2.5.4)
20. If gold bond wires are used within the packages, limit chip temperatures to 150°C max. (2.5.2)
21. Check case to ambient thermal gradients carefully, and ensure that individual devices are being tested at sufficiently high temperatures, or that adequate allowances are made for parametric degradations. (2.3.2, 2.5.5)
22. Include allowances for output loading and rail spikes when calculating package dissipations. (9.5)

Operating Rules

23. Do not connect any pin of a device to a low impedance voltage source more negative than −0·5 V with respect to the device earth. (4.1, 4.2)
24. Do not connect any input pin to a low impedance voltage source higher than 4·5 V above the device earth. (4.1)

25. Do not connect any low impedance positive voltage source to any device output pin. (4.2, 17.2)
26. Limit the supply rail voltage to 7 V absolute maximum, and make every effort to keep within the working limits quoted for the logic family in use. (17.1)
27. Any point in a T.T.L. circuit (other than the supply rail) can be earthed without risk of damaging the devices. (4.1, 4.2, 17.2.3)
28. Do not earth more than one output on any package at any time. (17.2.3)

Suggestions for Optimum use of Board Area

29. Use one sixteenth of an inch thick double-sided G10 board material.
30. Design board layouts on $X-Y$ co-ordinate principles.
31. Use a 0·050 in layout grid.
32. Use pads 0·050 in drilled 0·030 in plus or minus 0·002 in for all through-plated holes for mounting D.I.P.s or for via holes.
33. Use tracks 0·01875 in wide (0·075 in tape on 4 times artwork).
34. Run all tracks with their centres on the grid lines.
35. Make all bends in tracks at 45 degrees to the grid lines.
36. Start and finish all bends midway between grid lines.
37. Use a board connector with a pitch which is a multiple of 0·050 in.
38. Keep holes for D.I.P.s in rows on 0·300 in centres, i.e., mount 14- or 16-pin packages on 0·600 in centres side-to-side.

For further details on the above points on board layout, see *Printed Circuit Boards for Microelectronics*.

Appendix 2

Abbreviations used

a	Line held low by 10 Ω resistor at 'sending' end
b	Line held low by 10 Ω resistor at 'receiving' end
C	Capacitance
D.I.P.	Dual-in-line Package
D.T.L.	Diode-Transistor Logic
E	Earth
E	Line earthed at both ends
E	Amplitude of voltage swing on a line
E.C.L.	Emitter-Coupled Logic
E_r	Standing voltage at receiving end of a line
E_s	Standing voltage at sending end of a line
H	Line held high by a gate at sending end
h	Line held high by a gate at receiving end
I	Current
$I_{B\ IN}$	Inverse beta current
I_{CC}	Supply rail current
I_{IH}	Current into a device input in the '1' state ($=I_{IN\ LK}$)
I_{IL}	Current out of a device input in the '0' state
I_{OH}	Current out of a device output in the '1' state
I_{OL}	Current into a device output in the '0' state
I_{OS}	Current out of a '1' state device output when the output is shorted to earth
K_B	Back cross-talk constant
K_C	Track configuration constant
K_F	Forward cross-talk constant
KΩ	Kilo-ohms
L	Inductance
L	Line held low by a gate at sending end
l	Line held low by a gate at receiving end
l	Length of a line
l_{cr}	Critical length of a line
L.S.I.	Large Scale Integration
μF	Microfarads
mA	Milliamperes
M.S.I.	Medium Scale Integration
mW	Milliwatts

ns	Nanoseconds
P	Pick-up line
pF	Picofarads
Q	'data' output of flip-flop or M.S.I. element
$\overline{Q}$	'Not-data' output of flip-flop or M.S.I. element
R	Resistance
S	Signal Line
S	Logical sum output of a device
T	Delay time of a given line
t_p	Propagation delay along a line
t_{pd}	Half-pair-delay of a device—i.e. $(t_{PLH} + t_{PHL})/2$
t_{PHL}	Propagation delay of a device whose output falls in voltage
t_{PLH}	Propagation delay of a device whose output rises in voltage
t_r	Rise time of a pulse ($=t_{TLH}$)
t_{THL}	Fall time of device output (i.e., edge speed of falling voltage edge)
t_{TLH}	Rise time of device output (i.e., edge speed of rising voltage edge)
T.T.L.	Transistor-Transistor Logic
V	Voltage
V_B	Back cross-talk voltage
V_{BE}	Base-emitter voltage
V_{CC}	Supply rail voltage
V_{CE}	Collector–emitter voltage
V_F	Forward cross-talk voltage
$V_{IH\,min}$	Lowest '1' level voltage which a device input can accept as a '1' level
$V_{IL\,max}$	Highest '0' level voltage which a device input can accept as a '0' level
V_{OH}	'1' level output voltage of a device
V_{OL}	'0' level output voltage of a device
V_r	Instantaneous voltage at receiving end of a line
V_s	Instantaneous voltage at sending end of a line
V_{TH}	Input threshold voltage at which a device will switch
Z_o	Characteristic impedance of a line
Z_r	Impedance at receiving end of a line
Z_s	Impedance at sending end of a line
ρ_r	Coefficient of reflection at receiving end of a line
ρ_s	Coefficient of reflection at sending end of a line

Upper case letters are also used to define the inputs to a gate or circuit; the parts of a circuit; and successive line reflections.

Recommended Further Reading

MALEY, G. A. *Manual of Logic Circuits*. Prentice-Hall (1970)

Integrated Circuits—Design Principles and Fabrication, R. M. Warner and J. N. Fordemwalt (Eds). Motorola Series in Solid-state Electronics. McGraw-Hill (1965)

Analysis and Design of Integrated Circuits, D. K. Lynn, C. S. Meyer, D. J. Hamilton (Eds). Motorola Series in Solid-state Electronics. McGraw-Hill (1967)

SCARLETT, J. A. *Printed Circuit Boards for Microelectronics*, Van Nostrand Reinhold (1970)

'Proceedings of the Technical Programme', *International Electronic Packaging and Production Conference*, Industrial and Scientific Conference Management (1968, 69, 70)

Multilayer Printed Circuit Board Technical Manual, Institute of Printed Circuits

Application manuals, etc. published by the manufacturers of T.T.L. are not listed here, as these documents are liable to be updated and re-issued fairly frequently. All manufacturers will supply copies of their latest manuals, etc. on request.

Index

A.C. coupled flip-flop 122
A.C. noise margins 91ff, 109, 201
Acceleration 11
Adders 132
Alloying 9
Aluminium 6, 9, 18ff, 221
Ambient temperature (*see* Temperature)
AND function 23
AND gate 56, 67, 72, 99, 237
AND-OR-INVERT gate 25, 56, 65, 69, 80, 99, 118, 228, 237
Aperture (input switching) 1, 226
Application manuals 5
Arithmetic units (*see* Adders)

Back cross-talk 177 (*see also* Cross-talk)
Ball bonds (*see* Thermo-compression bonds)
Base charge 35, 37, 49
Beam leads 9, 19
Bistable 131, 227, 236, 240 (*see also* Flip-flop)
Black plague (*see* Plague)
Boards (*see* Printed circuit boards)
Bond wires 9, 18, 21, 22, 221
Bonding 9, 16 (*see also under* type of bond)
Bonding pads 6, 9, 155
Breakdown 39, 43
Buffers 62, 67, 72, 204, 222
Burn-in 17, 221

Calibration of testers 17
Capacitance
 line 157
 on output 2, 4, 47, 81, 90, 97, 104, 110, 207, 246
 stray 26, 47, 60, 169, 226
Capacitor
 decoupling (*see* Decoupling)
Ceramic 9, 11
Characteristic impedance (*see* Impedance)
Characteristics
 input 37, 169ff
 output 40, 169ff
 transfer 43, 56, 91
Charge (*see* Base charge)
Charge-storage flip-flop (*see* A.C. coupled flip-flop)
Chip 2, 8, 16, 150
Clock-skew 133, 148, 223
Co-efficients
 Back cross-talk K_B 177, 193, 203
 Forward cross-talk K_F 178
 Track configuration K_C 193, 203
Commercial devices 2, 11, 28, 65, 67, 74ff, 100
Comparators 132, 231
Complementary outputs 67, 70, 116, 131, 239
Computer-aided design 68, 236
Connectors 187, 205
Contact bounce 227
Contamination 2, 18
Cooling (*see* Thermal design)
Cost 2, 4, 126, 216, 230
Counters 131, 135, 223

Cracks 10, 16
Critical length 130, 162, 169, 199, 202
Cross-talk 73, 89, 110, 126, 156, 177ff, 187ff, 201ff, 215, 227, 231
Current density 20
Current spike (*see* Switching spike)
Current-steering logic (*see* Emitter-coupled logic)
Custom design 2, 11, 132

D-type flip-flop 116, 117, 121, 133ff
Data sheet (*see* Specification—manufacturer's)
Decoders 131
Decoupling 53, 208, 215ff
Defects 16
Design rules 73, 80, 248
Die (*see* Chip)
Dielectric constant 157, 180
Diffusion 6, 16, 37
Diode Transistor Logic 3, 4, 198, 228, 246
Diodes 3, 225
  input clamping 27, 36, 37, 176
  parasitic 6, 26, 39, 42, 176
Direct coupled flip-flop 118
Discontinuities 165ff, 187
Discrete component circuits 2, 5
Discrete wiring 198
Dissipation 1, 4, 21, 29ff, 51, 57, 64, 73, 96, 103, 104, 109, 110, 126, 148, 201, 218
Dual-in-line package 2, 8, 11, 155, 200

Earth return 5, 20, 212
Edge speeds 5, 74ff, 81, 83, 104, 125, 130, 162, 202, 223, 226
Edge-triggered flip-flop 120
Emitter-Coupled Logic 3, 4
Encoders 131
Engineering form 3, 4, 200
Epitaxial layer 6
Etching 6, 10, 16
Exclusive-OR 65, 70, 126, 168
Expanders 59ff, 67, 68, 72, 111
Extenders (*see* Expanders)

Fairchild 4, 30, 32
Fan-out 2, 28ff, 65, 67, 73ff, 84ff, 102, 104, 107, 110, 114, 124, 167, 172, 218, 228, 237, 239
Fault diagnosis 220, 231
Feedback 226
Ferranti 30
Flat Pack 2, 8
'Flip-chip' bonding 9
Flip-flop 11, 116ff, 126 (*see also under* type of flip-flop)
Forward cross-talk 178
Frequency 22, 57, 105

G.E.C. Semiconductors (*see* Marconi-Elliott Microelectronics)
Germanium 225
Gold 9, 18ff, 22, 37, 220
Graphical solution for reflections 161ff, 169, 180
Guard bands 11

Hybrid L.S.I. 2, 10

I.T.T. Semiconductors 30, 32
Impedances
  characteristic 157ff, 187ff, 201, 249
  input 37, 169ff
  output 1, 40, 169ff
  supply rail 53, 55, 205
  terminating 158, 222
  thermal 22
Impurities 18, 21
Inductance 110, 157, 205
Industrial devices (*see* Commercial devices)
Input current 37, 74ff, 84ff, 124, 127
Interconnections 5, 36, 60, 126, 156ff, 187ff, 218, 228
Interface specification 236
Inverse Beta Current 37
Ionic contamination 18
Isolation 6

J.K. flip-flop 116ff, 132, 237

L.S.I. 2 (*see also* Hybrid L.S.I.)
Labelling of devices 11
Laminate material 157
Lamp driver 63, 67, 72
Laser cutting 10
Latches 131
Lead frame 9, 155
Leakage current 21, 25, 37, 60, 74ff, 84, 169, 185
Length (of tracks) 110, 201ff
Line delay (*see* Propagation delay)
Line driver (*see* Buffer)
Load current 13, 22, 74ff, 84ff
Logic races 236, 246

M.O.S. 4, 132
Marconi-Elliott Microelectronics 30, 32
Marking of devices (*see* Labelling)
Matrix driving 226
Mean time to failure 22
Medium-scale integration 2, 8, 11, 22, 126ff, 200, 220
Memories 132
Metallization 2, 9, 11, 35, 126
Microstrip line 157ff, 187
Migration (metal) 19
Military devices 2, 11, 28, 65, 67
Milliwatt T.T.L. 34, 78, 82
Moisture 18
Motorola 29, 34
Mounting techniques (*see* Engineering form)
M.S. flip-flop 116, 135
Mullard 30
Multi-layer boards 53, 204, 213
Multiplexers 131

NAND gates 26, 56, 61, 65, 68, 80, 83, 99, 237
National Semiconductors 30
Noise 5, 60, 126, 135, 203
  immunity 5, 73, 89, 102, 104, 109, 110, 114, 150, 218
Noise margins 21, 84, 102, 201, 222
Nomenclature 65, 80

Non-inverting gate (*see* AND gate *or* OR gate)
NOR gate 59

OR function 24, 167, 228, 231, 237, 248
OR gate 59, 65, 72
Oscillation 226
Output current (*see* Load current)
Output voltage 13, 25, 29, 40, 74ff, 88, 107, 109
Overswing 161, 173ff, 202

Package allocations 60, 235
Pads (*see* Bonding pads)
Parallel working 4, 84, 218, 231
Parasitic
  capacitance (*see* Capacitance)
  diodes (*see* Diodes)
Phase-splitter 21, 23, 28, 56, 60, 106, 126
Philco-Ford 29
Photo-resist 6, 16
Pin allocations 31, 67
Plague 18
Planar process 3, 6
Planes (earth or power) 53, 157ff, 187, 201, 204, 213
Plastic encapsulation 8, 18
Platinum 19
Post-set time 121, 133, 149
Power
  consumption (*see* Dissipation)
  distribution 1, 5, 73, 83, 107, 204ff, 217
Pre-set time 121, 133, 149
Printed circuit boards 3, 5, 21, 29, 156, 178, 204ff, 229
Probe testing 11, 16
Propagation delay 74ff, 157, 159, 203, 223
P.T.F.E. 199
Pull-up circuit 1, 25, 42, 47ff, 69
Purchasing specification (*see* Specification)
Purple plague (*see* Plague)

Quality 16
control 13, 17

R.S. flip-flop 116
Race (*see* Logic race)
Raytheon 29, 33, 34, 69
Reflections 42, 51, 73, 89, 156ff, 187, 215, 222
Register 131, 228, 231
Relays 227
Reliability 3, 5, 10, 16, 18, 64, 126, 155, 221
Ringing (*see* Reflections)
Rules (*see* Design rules)

Schedule of interconnections 238
Schottky diode 35
Screening 5
Scribing 8
Series termination 222
S.G.S. 4, 30, 31
Shock 11
Shift register 131, 133, 223, 236
Silane 6
Silicon 2, 6, 16, 19
nitride 6
oxide (dioxide) 6
Silver 9
Size
of chips 2
of boards 200, 204, 230
Skew (*see* Clock-skew)
Slice (*see* Wafer)
Soldered joint 220
Spacing
of packages 200
of tracks 179, 202
Spares holding 231
Specifications
manufacturers 4, 16, 27, 74ff, 119, 124, 127, 207
purchasing 5, 16, 17, 119
Speed 1, 4, 11, 67, 73ff, 100, 104, 107, 110, 111, 156, 215, 225, 246
Sprague 30
Standard board layout 200
Stored base charge (*see* Base charge)
Substrate 6, 9
diodes (*see* Diodes)
Supply rail (*see* Power distribution)
Surges 84, 97, 221
Switches 227
Switching
spike 36, 49, 51, 53ff, 101, 103, 106, 110, 111, 115, 204, 219, 231
threshold 13, 37, 49, 102, 172, 183
Sylvania 4, 28, 30, 32, 150

Temperature
ambient 19, 21, 73
chip 11, 18, 19, 22, 97
storage 21
working 2, 11, 28, 80, 100
Terminations 157ff, 180, 189, 222
Testing
of boards 17, 219, 230
of devices 5, 11
goods inwards 17
Texas Instruments 4, 30, 33, 34, 35
Thermal
design 5, 21, 127
impedance (*see* Impedance)
Thermo-compression bonds 9, 18
Thickness of boards 158, 177, 194, 201, 202, 250
Through-plated holes 187, 220
Tin-lead 9, 220
Titanium 19
Tolerances 16, 27
Tracks
on chips 6
on boards 156, 165, 177, 187ff, 201ff
Transfer moulding 9
Transient control 57, 67, 72
Transistors 3
Transistor-coupled flip-flop 120
Transitron Electronic 4, 29, 30, 33, 69, 80, 81, 150
Transmission line 156ff, 187ff, 205
Tri-state output 2, 218
Type numbers 65, 127
Typical figures 5, 73, 82, 88, 119

Ultrasonic bonds 9, 18
Unrequired
  gates 99
  inputs 83, 84

Via holes 250
Vibration 11

Wafer 7, 10, 11
Westinghouse 29
Whiskers (on boards) 220
Width (of tracks) 19, 156ff, 165, 177, 191, 201, 250
Wired-OR 1, 60, 69, 167, 198, 218, 221, 228
Worst-case
  conditions 80
  design 73, 201, 207, 223